Introduction to Kinematics and Dynamics of Machinery

Synthesis Lectures on Mechanical Engineering

Synthesis Lectures on Mechanical Engineering series publishes 60–150 page publications pertaining to this diverse discipline of mechanical engineering. The series presents Lectures written for an audience of researchers, industry engineers, undergraduate and graduate students.

Additional Synthesis series will be developed covering key areas within mechanical engineering.

Introduction to Kinematics and Dynamics of Machinery
Cho W. S. To
2017

Solving Practical Engineering Mechanics Problems: Statics
Sayavur I. Bakhtiyarov
2017

Unmanned Aircraft Design: A Review of Fundamentals
Mohammad Sadraev
2017

Introduction to Refrigeration and Air Condition Systems: Theory and Applications
Allan Kirkpatrick
2017

Resistance Spot Welding: Fundamentals and Applications for the Automotive Industry
Menachem Kimchi and David H. Phillips
2017

MESMs Barometers Toward Vertical Position Detecton: Background Theory, System Prototyping, and Measurement Analysis
Dimosthenis E. Bolanakis
2017

Vehicle Suspension System Technology and Design
Avesta Goodarzi and Amir Khajepour
2017

Engineering Finite Element Analysis
Ramana M. Pidaparti
2017

Introduction to Kinematics and Dynamics of Machinery

Cho W. S. To

www.morganclaypool.com

ISBN: 9781681731759 paperback
ISBN: 9781681731872 ebook

DOI 10.2200/S00798ED1V01Y201709MEC007

A Publication in the Morgan & Claypool Publishers series
SYNTHESIS LECTURES ON MECHANICAL ENGINEERING

Lecture #7
Series ISSN
Print 2573-3168 Electronic 2573-3176

Introduction to Kinematics and Dynamics of Machinery

Cho W. S. To
Professor Emeritus
University of Nebraska, Lincoln

SYNTHESIS LECTURES ON MECHANICAL ENGINEERING #7

ABSTRACT

Introduction to Kinematics and Dynamics of Machinery is presented in lecture notes format and is suitable for a single-semester three credit hour course taken by juniors in an undergraduate degree program majoring in mechanical engineering. It is based on the lecture notes for a required course with a similar title given to junior (and occasionally senior) undergraduate students by the author in the Department of Mechanical Engineering at the University of Calgary from 1981 and since 1996 at the University of Nebraska, Lincoln. The emphasis is on fundamental concepts, theory, analysis, and design of mechanisms with applications. While it is aimed at junior undergraduates majoring in mechanical engineering, it is suitable for junior undergraduates in biological system engineering, aerospace engineering, construction management, and architectural engineering.

KEYWORDS

kinematics, dynamics, mechanisms, machinery, cams, gears, gear trains

Contents

Preface

There are many excellent books in *kinematics and dynamics of machinery*. The lengths of these books are generally more than 400 pages. It seems that there is a lack of shorter books in the range of 150 and 200 pages, so this book attemps to fill that gap. However, despite its shorter length, it still includes many examples with detailed solutions, and a sufficient number of topics that are normally covered in a required introductory course in kinematics and dynamics of machinery taken by junior undergraduates.

This book is based on the lecture notes that have been developed and used by the author since 1981. These lecture notes have been employed in the titled course first taught at The University of Calgary, Canada and subsequently at the University of Nebraska-Lincoln, U.S. It has primarily been taken by junior and some senior undergraduates with majors in mechanical engineering and biological system engineering. The subject matters dealt with in this book cover materials for a single-semester three-credit course. For a four-credit course, the component of laboratory experiments is omitted in the present shorter book for two main reasons. First, the book's smaller size makes inclusion of such experiments infeasible. Second, for scheduling purposes and availability of laboratory instrumentation in a department, laboratory experiments are better left to a separate course.

Under normal conditions, it is expected that the students using the present book have already taken sophomore-year courses in the following subjects: linear algebra, vector analysis, matrix theory, second course in mathematics, and engineering dynamics.

Cho W. S. To
December 2017

Acknowledgments

Chapters 2 and 3 were written with the assistance of Dr. Meilan Liu of Lakehead University, Canada. Her contribution to these chapters, and in particular her compilation of Table 3.7, is deeply appreciated by the author.

The author would like to express his sincere thanks to Paul Petralia, Executive Editor, Engineering, and his team members at Morgan & Claypool for their assistance and production effort on this book.

Cho W. S. To
December 2017

CHAPTER 1

Introduction

This book is concerned with the introduction to the kinematics and dynamics of machinery. The emphasis is on the fundamentals of kinematics of linkages along with the analytical and graphical methods for motion studies. Analytical studies and design of cams, commonly applied gears and gear trains, and dynamic forces are included in this book. In the present chapter, motivations, a brief historical highlights of development of machine elements and machinery, and organization of this book are included.

1.1 MOTIVATIONS

Before the main motivations of studying the principles of mechanism and machinery are addressed, the following basic definitions are presented.

Kinematics of machines   Is the a study of the relative motion of machine parts. That is, displacements, velocities, and accelerations are considered.

Dynamics of machines   Deals with the forces acting on the parts/components of a machine and the motions resulting from these forces.

What is a machine ?
  For the time being, the following simple definition is adopted:
A machine is a system that consists of a combination of bodies/elements arranged to compel mechanical forces to do work accompanied by motions.
  Mechanisms and machines have many applications and are encountered everywhere in the modern world. They pay a major role in the development of modern industries, such as mining, construction, automobile, shipping, railway, aeronautics, and aerospace. Thus, one main motivation of studying mechanisms and machinery is their role in the development of many industries. The second main motivation is to provide better designs and products of devices and engines, with better here means more durable, simple, and economical.

1.2 A BRIEF HISTORY

The history of the development of mechanisms and machines may loosely be divided into three phases: (1) the period before the industrial revolution; (2) the period during the industrial revolution; and (3) the current or post-industrial revolution period.

It is understood that there were mechanisms or inventions used before the industrial revolution. As far as mechanisms are concerned, the basic and important inventions, arguably, are the crank and gear-wheels. Of the crank and crank-like devices, it appears that the most ancient indubitable crank-handle was used in the Han period between 2nd century b.c. and 2nd century a.d. [1] . For three or four centuries before the time of Marco Polo, the powers of the crank were widely applied in China, employed in textile machinery for silk-reeling and hemp-spinning, in agriculture for the rotary winnowing-fan and water-powered flour-sifter, in metallurgy for the hydraulic blowing-engine, and in humbler applications such as the well-windlass. Gear-wheels dating back to around 230 BC have been found in tombs from the Qin through Han periods in China [1]. Another important invention that can be considered related to the gear-wheels is the device based on the reverse action to the modern differential gears commonly applied in automobiles. The differential gear, the so-called "nest of gear-wheels" used to be called the Starley gear, after the name of the inventor who first used it in vehicles in 1879 [1]. In fact, true differential gearing was introduced by Willamson around 1720 for correcting for the equation of time. The reverse action device to the differential gear has been traced back to the mechanism used in the south-pointing carriage designs [1].

In the second phase of the development of mechanisms during the industrial revolution period, many mechanical elements or devices were invented. The two authoritative books by Willis [2] and Reuleaux [3] contain many concepts and principles for such devices. In fact, Willis classified mechanism into elements of rolling contact, sliding contact, wrapping connectors, link-work, and reduplication. On the other hand, in his book Reuleaux dealt with pairs of elements, incomplete pairs of elements, incomplete kinematic chains, chamber-crank trains, chamber-wheel trains, constructive elements of machinery, complete machines, and kinematic synthesis.

In the third phase, the post-industrial or modern period, many mechanisms have been analyzed and designed with much improved theoretical foundation. During the late 1970s and early 1980s, robotic manipulators were developed with major applications in the automobile and aerospace industries. The sketches of a typical aircraft landing gear system, industrial robot, and a Wankel engine are shown in Figures 1.1–1.3, respectively. Theoretical development of mechanisms in three-dimensional (3D) space was received a considerable amount of effort during this period. For example, since the 1990s nano-mechanisms have been made. In addition, Newtonian mechanics-based mechanisms and machinery are ubiquitous in the present period. There are many excellent textbooks dealing with mechanisms and machinery, a few of which will be referenced in the next and subsequent chapters.

1.3   THE BOOK'S ORGANIZATION

The rest of the book is organized as follows.

- Chapter 2 deals with the fundamentals of kinematics, including the degrees-of-freedom (DOF), types of motion, links, joints, and kinematic chains.

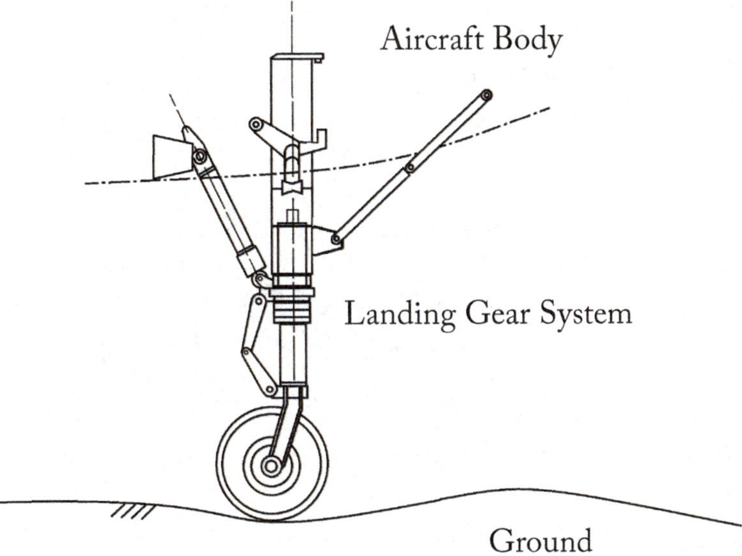

Figure 1.1: Plane view of a landing gear system.

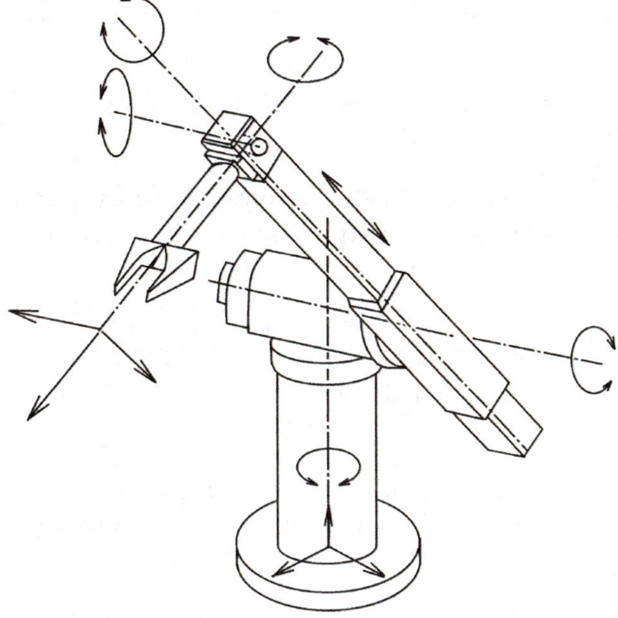

Figure 1.2: An industrial robot.

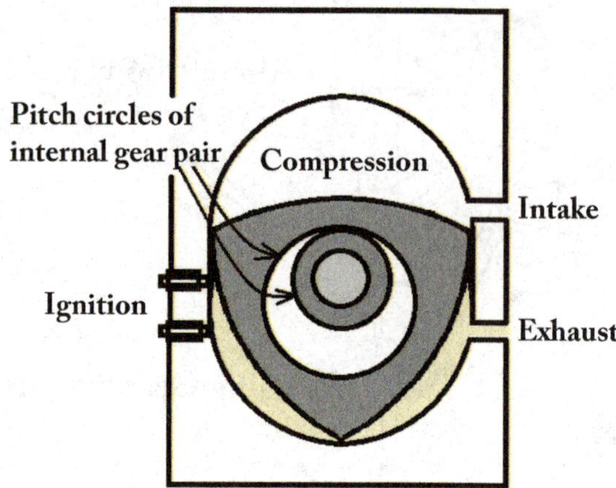

Figure 1.3: Cross-section view of a Wankel engine.

- The planar four-bar pin-jointed linkages are introduced in Chapter 3. The existence condition, Grashoff condition, inversions, transmission angles, toggle positions, and limits of rotations of input/output links are considered.

- In order to demonstrate the applications of the concepts presented in Chapter 2, one planar four-bar mechanism and one planar six-bar mechanism are included in Chapter 4.

- Chapter 5 deals with the analytical approaches in motion studies. The two methods in this chapter are the method of complex number and the unit vector or rotating frame of reference (RFR) solution method.

- Graphical approaches in motion studies are presented in Chapter 6. Two commonly applied methods considered in this chapter are the method of instantaneous centers and image method.

- Chapter 7 is concerned with the analysis and design of cams. The classifications of cams and followers, and analytical and graphical designs of cams are included. The analytical designs of a flat-faced follower and disk cam with oscillating roller follower are addressed in this chapter.

- Analysis and design of gears are introduced in Chapter 8. The emphasis is on standard spur gears and gear trains. In the latter, only fixed gear centers and planetary gear trains are studied.

- Analysis of dynamic forces in machinery is presented in Chapter 9. To limit the scope of the present book, only the D'Alembert's principle, method of virtual work, and gyroscopic forces are considered in this final chapter.

REFERENCES

[1] Needham, J., *Science and Civilization in China*, Volume 4, Part 2. Cambridge, Cambridge University Press, 1965. 2

[2] Willis, R., *Principles of Mechanism*. Cambridge, Cambridge University Press, 1870. 2

[3] Reuleaux, F. (Translated by Kennedy, A. B. W.), *The Kinematics of Machinery: Outlines of a Theory of Machines*. New York, MacMillan and Company, 1876. 2

C H A P T E R   2

Fundamentals of Kinematics

Before the analysis of motion in mechanisms and machines, the fundamentals of kinematics of machinery are considered in this chapter. These are the degrees-of-freedom (DOF), types of motion, links, joints, and kinematic chains. Section 2.1 deals with the DOF of mechanisms. Types of motion are introduced in Section 2.2 while links, joints, and kinematic chains are presented in Section 2.3. The final section, Section 2.4, is concerned with the determination of DOF of mechanisms.

2.1 DEGREES-OF-FREEDOM (DOF)

The DOF of a system is the number of *independent* parameters (measurements) which are needed to *uniquely* describe its position in space at any instant in time.

- DOF are defined with respect to (w.r.t.) a selected frame of reference (FR). Frequently, one chooses the two-dimensional (2D) FR for planar motion or 3D FR for spatial motion.

- For a given system, the number of *independent* parameters (measurements) does not change with the selection of FR, but the parameters themselves are not unique.

Example 2.1
For planar motion in 2D space, every unconstrained rigid body has three DOF. The unconstrained rigid rod in the 2D Cartesian co-ordinate system is shown in Figure 2.1 in which the FR has the X and Y co-ordinates, and *independent* parameters are: x_A, y_A, and θ. That is, in the plane motion the unconstrained rigid rod can move independently along the X-axis, Y-axis, and rotates about an axis perpendicular to the plane that contains the X and Y co-ordinates.

Example 2.2
For motion in 3D space, every unconstrained rigid body has six DOF. The unconstrained rigid rod in the 3D Cartesian co-ordinate system is shown in Figure 2.2 in which the FR has the X, Y, and Z co-ordinates, and *independent* parameters are: x_A, y_A, z_A, θ, φ, and γ where, for example, θ is the angle between the rigid rod AB and the X-axis. That is, in the 3D space the unconstrained rigid rod can translate independently along the X-axis, Y-axis, and Z-axis, and rotates independently with angles θ, φ, and γ.

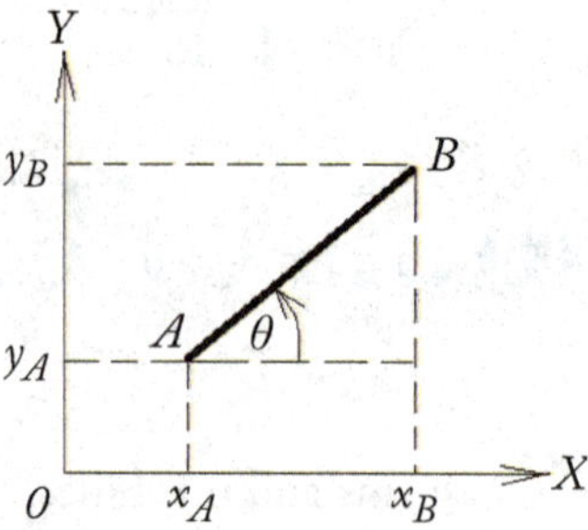

Figure 2.1: Unconstrained rigid rod in 2D space.

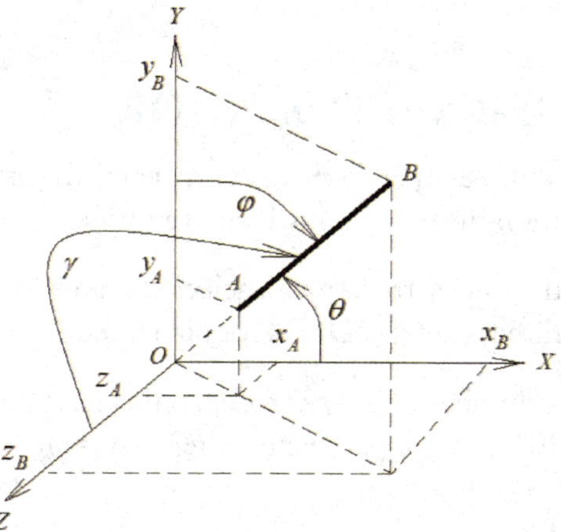

Figure 2.2: Unconstrained rigid rod in 3D space.

2.2   TYPES OF MOTION

For conciseness only motions in 2D space are considered. Three different types of motion can be identified. They are listed in the following.

(a) **Pure rotation**

The rigid body possesses one point which has no motion with respect to the "stationary" FR. The latter point is the center of rotation. All other points on the body will describe arcs about the center of rotation. A reference line drawn on the body through the center changes only its angular orientation.

(b) **Pure translation**

All points on the body describe parallel (rectilinear or curvilinear) paths. A reference line drawn on the body changes its linear position, but does not change its angular orientation.

(c) **Complex motion**

Complex motion is a simultaneous combination of rotation and translation. Any reference line drawn on the body will change its linear position and angular orientation. Points on the body will travel non-parallel paths, and there will be, at any time instant, a center of rotation which will continuously change location.

For illustration, if the three *independent* parameters are x_A, y_A, and θ then

for pure rotation: x_A and y_A are constant, and $\theta = \theta(t)$ in which A is the center of rotation;

for pure translation: $\theta = $ constant, $x_A = x_A(t)$, $y_A = y_A(t)$; and

for complex motion: $x_A = x_A(t)$, $y_A = y_A(t)$, $\theta = \theta(t)$.

2.3   LINKS, JOINTS, AND KINEMATIC CHAINS

In this section, definitions and various types of joints are presented. The definitions are given in the following first.

Node: a point at which a rigid body is attached to the other(s).

Link: a rigid body which possesses at least two nodes. (Of course, in general, flexible links can be included in a mechanism. However, for simplicity, flexible links are not considered in this book.)

Joint: a connection between two or more links at their nodes; such connection allows motions between the connected links.

Kinematic chain: an assembly of links and joints, inter-connected in a way to provide a controlled output motion in response to an input motion.

Mechanism: (or linkage) a kinematic chain in which at least one link has been "grounded," or attached to the frame of reference.

Machine: a combination of bodies arranged to compel the mechanical forces to do work accompanied by motion(s).

Different Types of Joints.   Various types of joints are included in the following.

(a) *One DOF joints*

These joints allow only rotation or translation. An example of the rotation type is the pin joint and example for the translation type is the slider joint. These joints are called full joints. They are also referred to as *lower pairs* (if two links make contact over a surface the pair is called a *lower pair*).

(b) *Two DOF joints*

These joints allow both rotation and translation. An example of this type is the pin-in-slot joint. These joints are called half joints. They are also referred to as *higher* or *upper pairs* (when two links make contact along a line or at a point the pair is called a *higher pair*).

(c) *Three DOF joints*

These joints allow rotation about three axes. A typical example is the ball and socket joint.

(d) *Multiple joints*

These joints involve more than two links being pinned at the same point. The number of one DOF joints are obtained by the following relation:

Number of one DOF joints = Number of links pinned at the same point -1.

2.4   DETERMINATION OF DOF

The definition of DOF has been provided in Section 2.1. In a planar mechanism a commonly adopted equation for the determination of DOF is given by Kutzbach [1]

$$\lambda = M = 3(L - 1) - 2J_1 - J_2,\tag{2.1}$$

where

$$
\begin{aligned}
L &= \text{number of links (including the ground),}\\
J_1 &= \text{number of full (1 DOF) joints,}\\
J_2 &= \text{number of half (2 DOF) joints,}\\
\lambda &= \text{number of DOF of the linkage or mechanism, and}\\
M &= \text{number of degrees of mobility (DOM).}
\end{aligned}
$$

As far as planar motion is concerned and in the U.S. no difference is made between λ and M, for example, $\lambda = M$ is adopted in [1, 2, 3, 4]. However, in 3D motion and in the European school the concepts of DOF and DOM are distinguished [5]. In this book Equation (2.1) is employed in the determination of DOF for planar mechanisms.

Explanation of Equation (2.1): In a 2D space there are three DOF for every unconstrained rigid body. If one link is grounded then $\lambda = 3(L - 1)$. A one DOF joint removes two DOF (since it cannot move in the X and Y co-ordinates, but can rotate). A two DOF joint removes one DOF, and hence,

$$\lambda = M = 3(L - 1) - 2J_1 - J_2.$$

2.4.1   REMARKS

In applying Equation (2.1) for the determination of DOF of mechanisms the following remarks in regard to gear contact, spring, as well as belt and pulley (or chain and sprocket) are in order.

(a) *Gear Contact*

In gear systems, one should consider both roll and slide contact between gear teeth. Thus, the contact point (so-called pitch point) is considered a two DOF joint. In practice, when two gears are in mesh they are rolling and sliding because of friction and therefore gear contact is considered a two DOF joint, unless it is stated otherwise.

(b) *Spring*

In a spring there is no movement constraint and therefore it is disregarded in the computation of DOF.

(c) *Belt and Pulley (Chain and Sprocket)*

In these elements only the tension side of belt/chain is considered as a link. Points of tangency are considered as full (1 DOF) joints (meaning there is no-sliding between belt/pulley or chain/sprocket for practical reasons since no power will be available if sliding occurs).

2.4.2   PARADOXES

It is important to note that Kutzbach's equation may lead to false results! This is because it does not mention the size of links or over-constrained links. Mathematically, Kutzbach's equation just gives the necessary condition. The sufficient condition is the *existence condition* (to be introduced in Chapter 3).

(a) *Over-constrained Linkages*

These linkages seem to have special geometry. For example, in Figure 2.3 there are five links and six one DOF joints. Therefore, according to Equation (2.1),

$$\lambda = 3(5 - 1) - 2(6) - 0 = 0.$$

This means the system in Figure 2.3 is a structure since there is no DOF. However, it should be $\lambda = 1$.

Another example is a pair of gears in mesh which is shown in Figure 2.4.

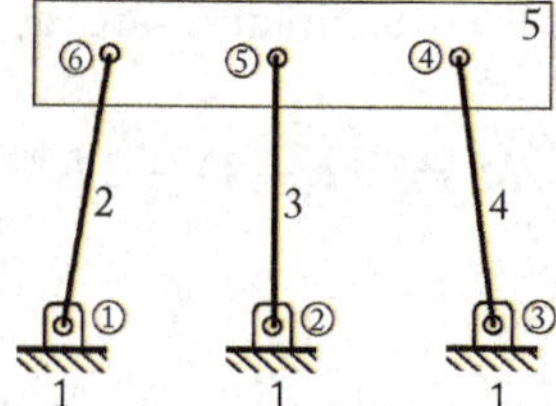

Figure 2.3: Over-constrained linkage: integer denotes link number and circled integer designates full-joint number.

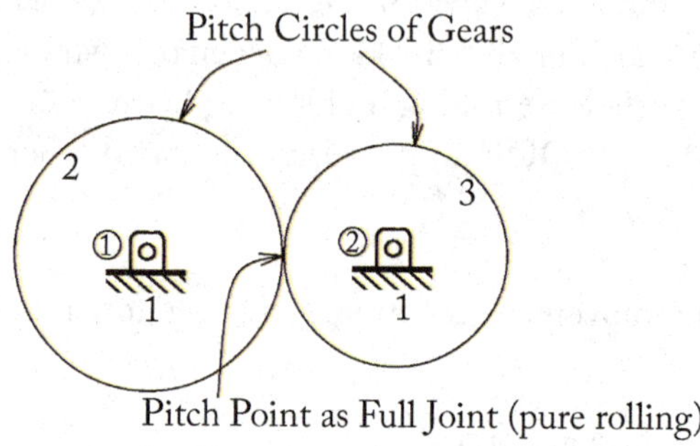

Figure 2.4: Two gears in mesh.

The above linkage has: $L = 3$, $J_1 = 3$, $J_2 = 0$. Hence,

$$\lambda = 3(3 - 1) - 2(3) - 0 = 0.$$

In practice, the gear pair rotates and therefore it is $\lambda = 1$. In other words, there should be friction that ensures sliding as well as rolling so that the pitch point is a two DOF joint. Then, $L = 3$, $J_1 = 2$, $J_2 = 1$ such that $\lambda = 3(3 - 1) - 2(2) - 1 = 1$.

(b) *Redundant DOF*

When rollers are employed to reduce friction but they do not affect the motion overall. The cam with a roller follower shown in Figure 2.5 is a good example.

In Figure 2.5, $L = 4$, $J_1 = 3$, $J_2 = 1$. Hence, $\lambda = 3(4 - 1) - 2(3) - 1 = 2$. However, the rotation of link 4 (roller) does not affect the oscillatory rotation of link 3 (follower). There is one redundant DOF. The correct interpretation should be: $L = 3$, $J_1 = 2$, $J_2 = 1$ so that $\lambda = 3(3 - 1) - 2(2) - 1 = 1$.

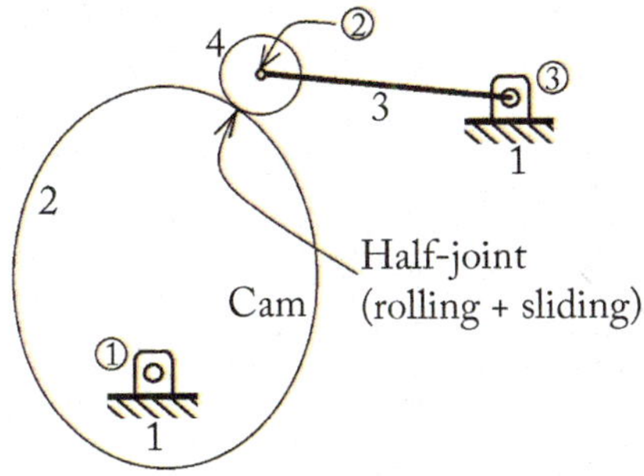

Figure 2.5: Cam with a roller follower.

2.4.3 EXAMPLES

The following four examples are included to illustrate the determination of the DOF of the mechanisms.

Example 2.3
Consider the four-bar pin-jointed linkage shown in Figure 2.6. Determine λ for this linkage.

Solution: $L = 4, J_1 = 4, J_2 = 0$. Hence, $\lambda = 3(4 - 1) - 2(4) - 0 = 1$.

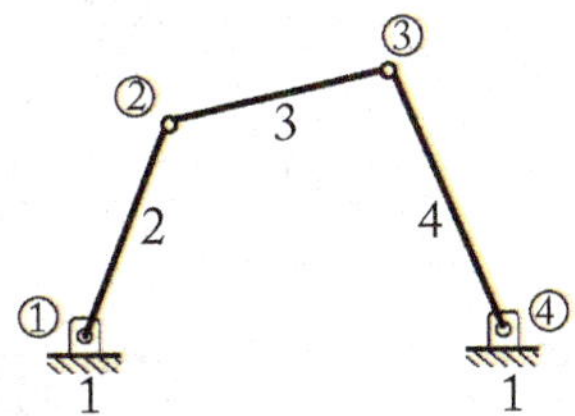

Figure 2.6: A four-bar mechanism.

Example 2.4
Consider the linkage shown in Figure 2.7. Determine λ.

Solution: $L = 8, J_1 = 10, J_2 = 0$. Hence, $\lambda = 3(8 - 1) - 2(10) - 0 = 21 - 20 = 1$, or $L = 7, J_1 = 8, J_2 = 1$ so that $\lambda = 3(7 - 1) - 2(8) - 1 = 1$. Note that the two different interpretations of the linkage give the same λ.

Example 2.5
Consider the system shown in Figure 2.8. Determine λ.

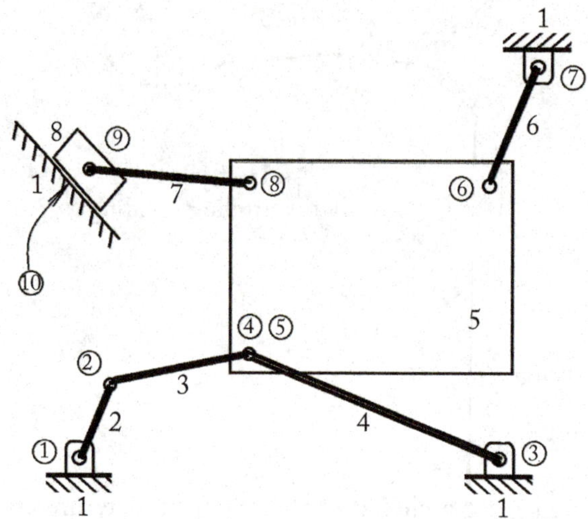

Figure 2.7: An eight-bar mechanism.

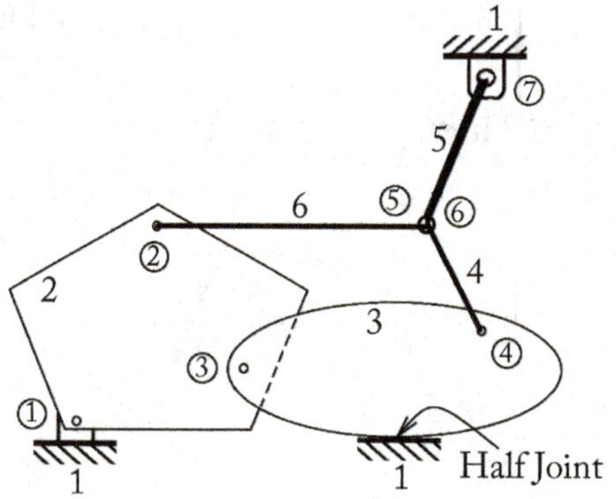

Figure 2.8: A six-bar mechanism.

Solution: $L = 6$, $J_1 = 7$, $J_2 = 1$. Hence, $\lambda = 3(6-1) - 2(7) - 1 = 0$. This means the given system is a structure.

Example 2.6

Consider the linkage in Figure 2.9. Applying Kutzbach's equation determine λ.

Solution: $L = 7$, $J_1 = 7$, $J_2 = 1$. Hence, $\lambda = 3(7-1) - 2(7) - 1 = 3$ or $L = 6$, $J_1 = 5$, $J_2 = 2$. Hence, $\lambda = 3(6-1) - 2(5) - 2 = 3$.

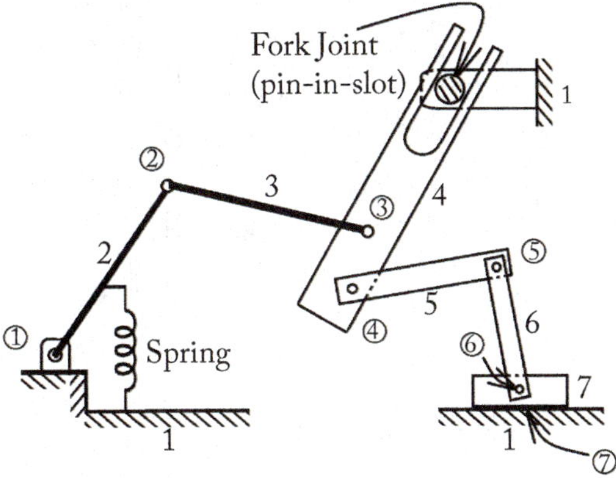

Figure 2.9: A seven-bar mechanism with a spring.

2.5   EXERCISES

2.1. A control rod passes through a horizontal hole in the body of the toggle clamp, as shown in Figure 2.10. Draw a kinematic diagram of this mechanism and determine its DOF.

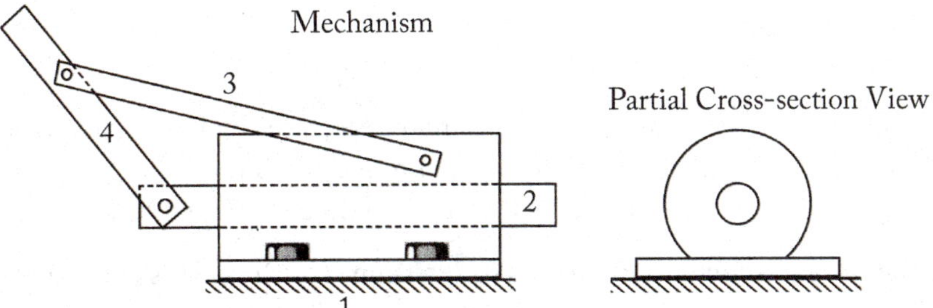

Figure 2.10: Toggle clamp.

2.2. An automobile front-wheel assembly is shown in Figure 2.11. Draw a kinematic sketch and determine its DOF. Note that since the front-wheel assembly contains two symmetrical sides only one side is being shown in Figure 2.11.

2.3. The linkage shown in Figure 2.12 has a roller-in-slot at B while a cylindrical roller at E supports the horizontal rigid beam DE. The cylindrical roller at E is perpendicular to the plane of the figure and it is always in contact with the horizontal rigid beam. Draw a kinematic sketch of the linkage shown and determine its DOF.

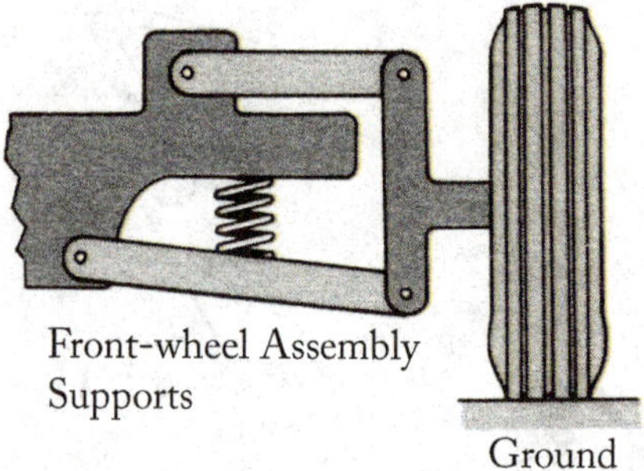

Figure 2.11: An automobile front-wheel assembly.

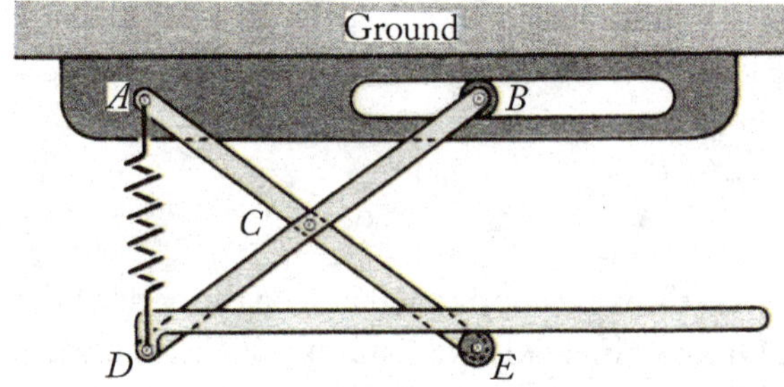

Figure 2.12: A mechanism with a roller-in-slot.

2.4. A mobile overhead working platform is mounted on the bed of a truck. It consists of two identical mechanisms symmetrically mounted. Only one of the mechanisms is shown in Figure 2.13. Draw a kinematic diagram and determine its DOF.

2.5. For the linkage shown in Figure 2.14, determine its DOF. It is assumed that all sliders B, D, and E are always in contact with the ground.

2.6. For the linkage shown in Figure 2.15, determine its DOF.

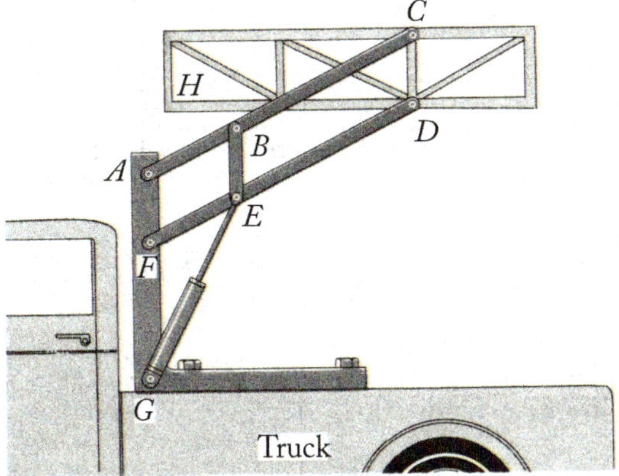

Figure 2.13: A mobile overhead working platform.

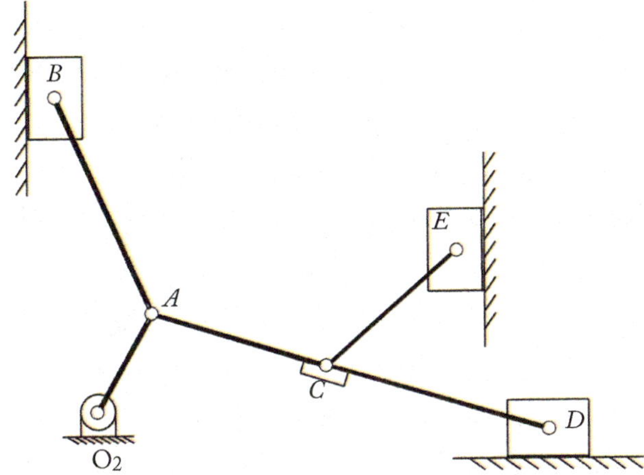

Figure 2.14: A mechanism with three sliders.

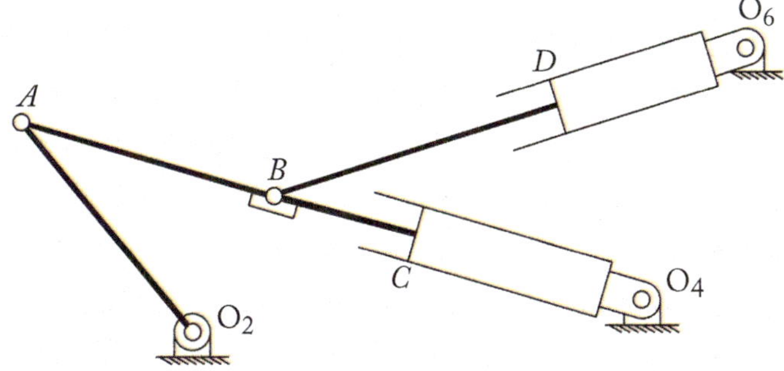

Figure 2.15: A mechanism with two actuators.

REFERENCES

[1] Norton, R. L., *Design of Machinery*, 2nd ed., New York, McGraw-Hill, 1999. 10

[2] Paul, B., *Kinematics and Dynamics of Planar Machinery*, Englewood Cliffs, NJ, Prentice-Hall, 1979. 10

[3] Erdman, A. G. and Sandor, G. N., *Mechanism Design: Analysis and Synthesis*, 3rd ed., Upper Saddle River, NJ, Prentice-Hall, 1997. 10

[4] Myszka, D. H., *Machines and Mechanisms: Applied Kinematic Analysis*, 2nd ed., Upper Saddle River, NJ, Prentice-Hall, 2002. 10

[5] N-Nagy, F. and Siegler, A., *Engineering Foundations of Robotics*, Englewood Cliffs, NJ, Prentice-Hall International, 1987. 10

CHAPTER 3

Planar Four-Bar Pin-Jointed Linkages

In the analysis and design of mechanisms, planar four-bar pin-jointed or simply four-bar mechanisms belong to a class for its simplicity and many applications. In this chapter various important concepts for four-bar mechanisms are presented. Section 3.1 introduces the existence condition. Section 3.2 deals with the Grashof condition. Inversions and transmission angles are included in Sections 3.3 and 3.4, respectively. Toggle positions are considered in Section 3.5. Circuits are introduced in Section 3.6 and limits of rotations of input/output links are presented in Section 3.7. Summary is included in Section 3.8.

Before presenting the aforementioned concepts the main advantages of studying four-bar mechanisms and terminology are included first.

The main advantages of four-bar mechanisms are:

(a) it has one single DOF and therefore it requires only one kinematic input;

(b) simplicity since it contains the smallest number of links in a linkage; and

(c) the basis of analysis as more complex linkages can be divided into several sub-four-bar linkages.

Consider Figure 3.1 in which a four-bar mechanism is presented.

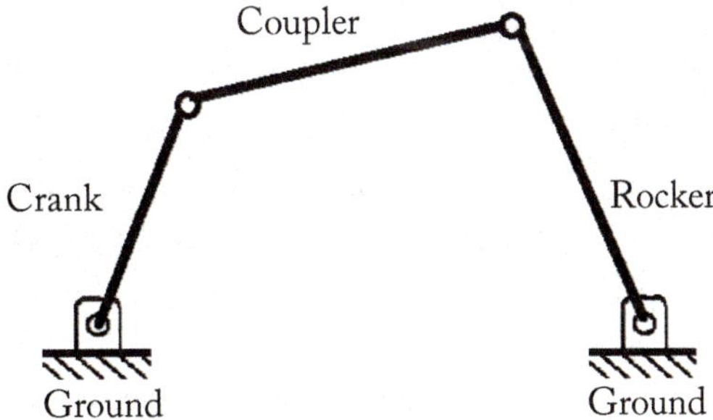

Figure 3.1: Planar four-bar pin-jointed mechanism.

Terminology.

Ground: a link (links) which is (are) fixed with respect to the frame of reference (that is, inertia frame or Newtonian frame).

Crank: a link which makes a complete revolution and is pivoted to the ground.

Coupler: a link that has complex motion and is not pivoted to the ground.

Rocker: a link that has oscillatory rotation and is pivoted to the ground.

3.1   EXISTENCE CONDITION

Let S be the shortest link (in terms of length), L the longest link (not to be confused with the number of links in Chapter 2), and P and Q the intermediate links.

If a four-bar linkage exists then the following relation is true:

$$L \leq S + P + Q. \tag{3.1}$$

Three cases are possible. The first case is

$$L > S + P + Q$$

in which the links cannot be assembled together.

The second case is

$$L = S + P + Q$$

in which the links can be assembled together, but cannot be moved as a linkage.

The third case is

$$L < S + P + Q$$

in which the links can be assembled together and can be moved as a linkage.

These three cases are illustrated in Figure 3.2.

3.2   GRASHOF CONDITION

- In a planar four-bar pin-jointed linkage the *Grashof condition* indicates whether or not at least one of the links can perform a full revolution.

- It does not determine which link being able to make the complete revolution. This is to be answered by inversion (see later). For a four-bar pin-jointed linkage, if

$$S + L \leq P + Q \tag{3.2}$$

is satisfied, the linkage is said to be *Grashof* which means that at least one link is capable of making a full revolution with respect to the ground plane.

If Equation (3.2) is not satisfied, the linkage is *non-Grashof* and no link is capable of performing a complete revolution with respect to the ground plane.

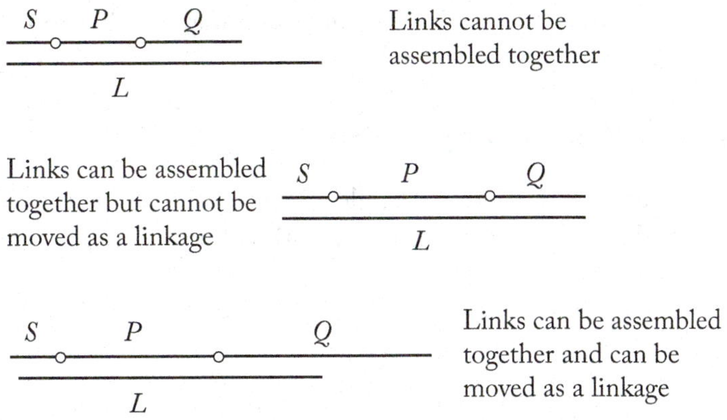

Figure 3.2: Three possible cases of existence and non-existence condition.

3.3   INVERSION

- Inversion is grounding a different link in a linkage.

- There are as many inversions of a given linkage as its links.

The motions resulting from each inversion can be quite different, but some inversions may yield motions similar to other inversions of the same linkage. The term *distinct inversions* is used to denote the inversions which have *distinctly* different motions.

Example 3.1
Consider a Grashof four-bar pin-jointed linkage that satisfies $S + L < P + Q$. How many distinct motions are there for this linkage?

Solution: There are three distinct inversions.

(a) Grounding either link adjacent to the shortest link implies the linkage is a crank-rocker. The shortest link is the crank.

(b) Grounding the shortest link implies the linkage is a double-crank. The links adjacent to the shortest link are the cranks.

(c) Grounding the link opposite to the shortest link implies the linkage is a Grashof double-rocker. Thus, only the coupler can make a whole revolution.

Example 3.2
Consider a non-Grashof four-bar pin-jointed linkage that satisfies $S + L > P + Q$. How many distinct motions are there for this linkage?

Solution: There is one distinct inversion since by grounding any link one has non-Grashof double-rockers.

3.4   TRANSMISSION ANGLES

In this section the definition of transmission angle, its computation, and the reason of its consideration in the context of four-bar mechanisms are outlined, respectively, in the following three sections.

3.4.1   DEFINITION

Consider a four-bar pin-jointed linkage shown in Figure 3.3.

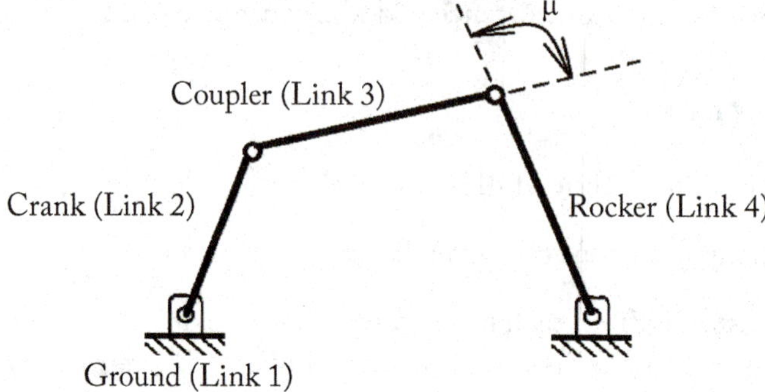

Figure 3.3: Transmission angle in a four-bar mechanism.

With reference to Figure 3.3, transmission angle μ is defined as the angle between the output link and the coupler. It is expressed in absolute value and as an acute angle.

For example, if it is found by computation that $\mu = 105°$ then the expressed value of $\mu = 180° - 105° = 65°$ is used.

To compute the transmission angle the *laws of cosines* are employed. Recall, for a triangle such as the one in Figure 3.4, the angle α is computed by using the following *law of cosine*: $a^2 = b^2 + c^2 - 2bc \cos \alpha$ so that

$$\cos \alpha = \frac{b^2 + c^2 - a^2}{2bc}, \tag{3.3}$$

where a, b, and c are the lengths of the sides of triangle shown in Figure 3.4.

3.4.2   DETERMINATION OF μ

In motion analysis and design, the transmission angle μ varies continuously from some minimum to some maximum value as the linkage undergoes its range of motion. In order to provide

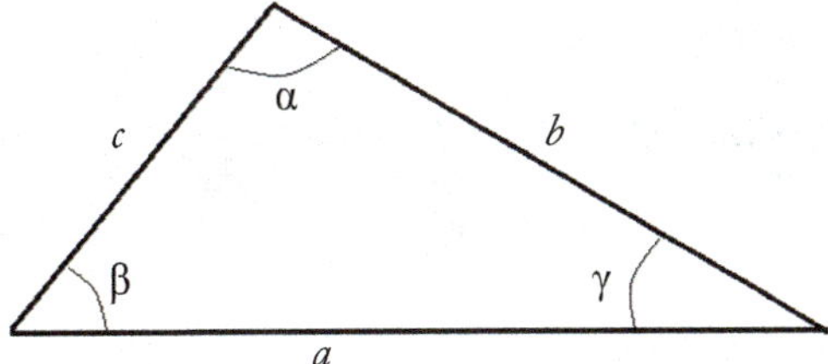

Figure 3.4: A triangle for using laws of cosines.

a better understanding so as to be able to determine μ, the following types of mechanisms are considered.

(a) **_Grashof crank-rocker linkages._** In this type of linkages μ is between a minimum value $\mu_{\min}$ and maximum value $\mu_{\max}$. That is,

$$\mu_{\min} \leq \mu \leq \mu_{\max},$$

where

$$\mu_{\min} = \min\{\mu_1, \mu_2\}, \quad \mu_{\max} = \max\{\mu_1, \mu_2\},$$

and μ_1 and μ_2 are defined in Table 3.1.

(b) **_Grashof rocker-crank, double-crank, and double-rocker linkages._** In these three types of linkages μ runs between zero and $90°$. That is,

$$0° \leq \mu \leq 90°.$$

(c) **_Non-Grashof double-rocker linkages._** For this type of linkages μ moves between zero and a maximum value. Thus,

$$0° \leq \mu \leq \mu_{\max},$$

where $\mu_{\max}$ is given in Table 3.2.

3.4.3   REASON FOR STUDYING TRANSMISSION ANGLES

Transmission angle is a measure of the quality of force and velocity transmission at the joint. Therefore, the issues of force and velocity transmissions are addressed.

(a) **_Force transmission._** Consider the four-bar mechanism in Figure 3.5 in which $\vec{F}_{34}$ is the force acting on link 4 caused by link 3. The tangential component of $\vec{F}_{34}$ is $\vec{F}_{34}^{\,t}$. Thus, the magnitude of the latter force is

$$F_{34}^{t} = F_{34} \sin\mu.$$

Table 3.1: Limiting values of μ for Grashof crank-rocker linkages

a = crank, b = coupler	c = rocker, d = ground
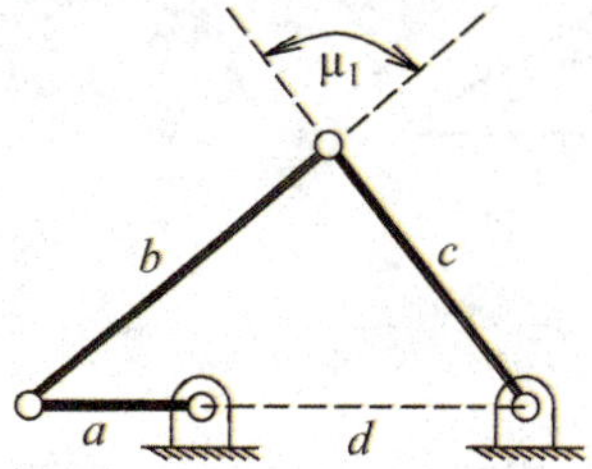	$\mu_1 = \cos^{-1}\left[\dfrac{b^2 + c^2 - (d+a)^2}{2bc}\right]$
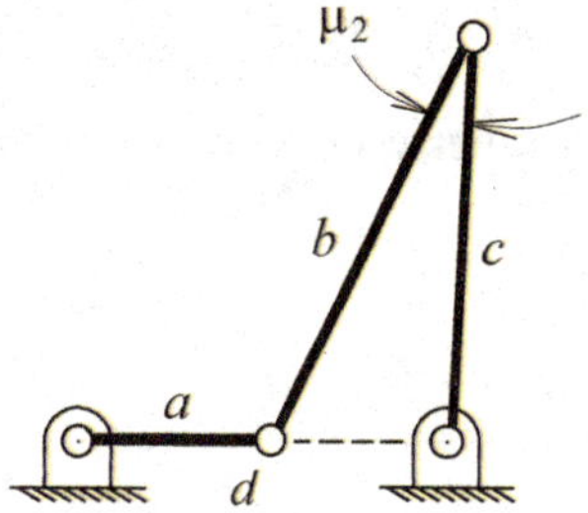	$\mu_2 = \cos^{-1}\left[\dfrac{b^2 + c^2 - (d-a)^2}{2bc}\right]$

Table 3.2: Maximum values $\mu_{\max}$ for non-Grashof double-rocker linkages

a = crank, b = coupler	c = rocker, d = ground
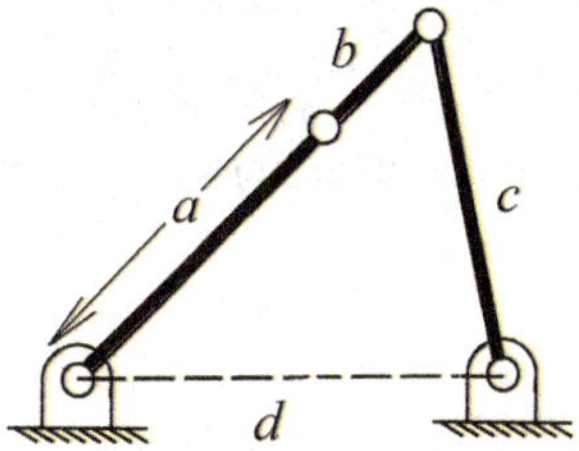	$\mu_{\max} = \cos^{-1}\left[\dfrac{(a+b)^2 + c^2 - d^2}{2c\,(a+b)}\right]$
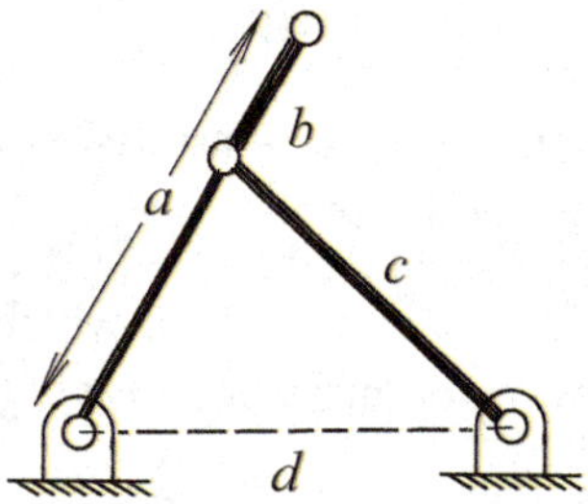	$\mu_{\max} = \cos^{-1}\left[\dfrac{(a-b)^2 + c^2 - d^2}{2c\,(a-b)}\right]$

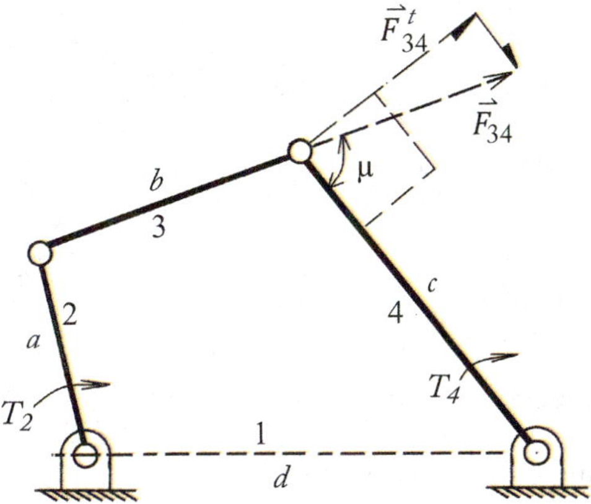

Figure 3.5: Force transmission.

The corresponding couple is

$$T_4 = c F_{34}^t = c F_{34} \sin \mu.$$

This means that increasing μ corresponds to increasing T_4. The optimal value

$$(T_4)_{\max} = c F_{34}^t \quad \text{when} \quad \mu = 90°.$$

(b) **Velocity transmission.** Consider Figure 3.6. The *principle of transmissibility* states that velocity at point A can be transmitted to point B. That is,

$$\omega_4 c \sin \mu = \omega_2 a \sin \nu$$

which gives the velocity ratio as

$$\varphi = \frac{\omega_4}{\omega_2} = \frac{a \sin \nu}{c \sin \mu}, \tag{3.4}$$

where ν is the angle between the input link and the coupler.

From Equation (3.4), it is observed that as the transmission angle increases the velocity ratio decreases. The velocity ratio approaches infinity when the transmission angle approaches zero.

(c) **Summary.** It is clear that for better force transmission the transmission angle μ should be increased. On the other hand, for better velocity transmission the transmission angle μ should be reduced. Therefore, in most machine design it is recommended to keep $\mu > \mu_{\min} = 35°$. The optimal value for μ is $90°$.

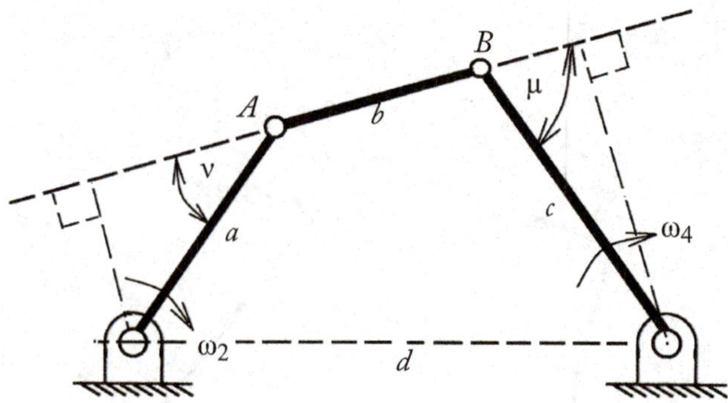

Figure 3.6: Transmission angle in a four-bar mechanism.

3.5   TOGGLE POSITIONS

- Toggle positions are determined by the *co-linearity of any two moving links*. Thus, the co-linearity can be between input link and coupler, and coupler and output link. Toggle positions for various types of linkages are considered in Section 3.5.1 and the reason for studying toggle positions is provided in Section 3.5.2.

3.5.1   TOGGLE POSITIONS

Toggle positions occur in Grashof crank-rocker, rocker-crank, and double-rocker linkages. In non-Grashof linkages only double-rocker type has toggle positions. These toggle positions are considered in the following.

(a) **Grashof crank-rocker linkages.** In this type of linkages co-linearity between coupler and rocker is *unobtainable*. Co-linearity between crank and coupler is obtainable with two possibilities: *extended* co-linearity and *overlapping* co-linearity.

These toggle positions are included in Table 3.3. Note, for Grashof rocker-crank linkages the toggle positions are similar except that now co-linearities are between coupler and output link (the input link is the rocker and the output link is the crank).

(b) **Grashof double-rocker linkages.** There are four toggle positions in this type of linkages. Toggles occur in combinations of any two moving links that can have both extended and overlapping co-linearities. These toggle positions are presented in Table 3.4. Note that Grashof double-crank linkages do not have toggle position.

(c) **Non-Grashof double-rocker linkages.** There are two toggle positions in this type of linkages. Toggles occur in combinations of any two moving links that can only have either extended or overlapping co-linearities. These toggle positions are presented in Table 3.5.

Table 3.3: Toggle positions for Grashof crank-rocker linkages

a = crank, b = coupler	c = rocker, d = ground
	$\theta_2 = \cos^{-1}\left[\dfrac{(a+b)^2 + d^2 - c^2}{2d\,(a+b)}\right]$
	$\theta_2 = -\cos^{-1}\left[\dfrac{(a-b)^2 + d^2 - c^2}{2d\,(a-b)}\right]$

3.5.2   REASON FOR STUDYING TOGGLE POSITIONS

The reason for studying toggle positions is given in the following.

(a) ***Toggle and velocity transmission.*** Recall that the velocity ratio defined in Equation (3.4) is

$$\varphi = \frac{\omega_4}{\omega_2} = \frac{a \sin \nu}{c \sin \mu}. \tag{3.4}$$

When the angle ν approaches zero ω_4 also approaches zero, regardless of ω_2.

When the angle μ approaches zero ω_2 also approaches zero, regardless of ω_4. For the latter case, see, getting out of toggle later in this section.

(b) ***Toggle and force transmission.*** Recall that in dealing with transmission angles in the foregoing Section 3.4.3 one has the force transmission relations

$$F_{34}^t = F_{34} \sin \mu, \qquad T_4 = cF_{34}^t = cF_{34} \sin \mu.$$

The implication of the last equation is that:

when the angle μ approaches zero T_4 also approaches zero, regardless of T_2;

when the angle $\mu = 90°$ the output torque $(T_4)_{\max} = cF_{34}$.

Table 3.4: Toggle positions for Grashof double-rocker linkages

a = crank, *b* = coupler	*c* = rocker, *d* = ground
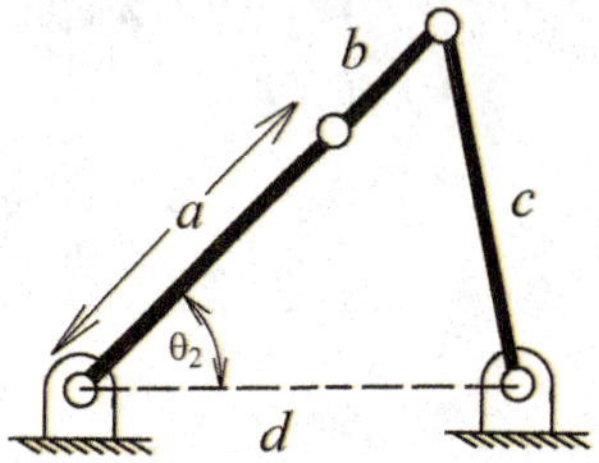	$\theta_2 = \cos^{-1}\left[\dfrac{(a+b)^2 + d^2 - c^2}{2d\,(a+b)}\right]$
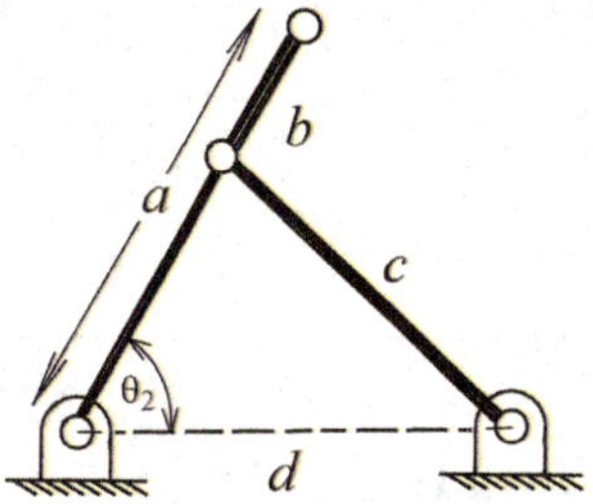	$\theta_2 = \cos^{-1}\left[\dfrac{(a-b)^2 + d^2 - c^2}{2d\,(a-b)}\right]$
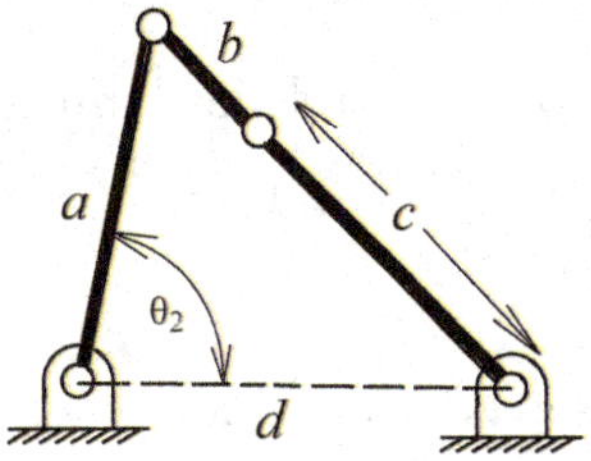	$\theta_2 = \cos^{-1}\left[\dfrac{d^2 + a^2 - (b+c)^2}{2ad}\right]$
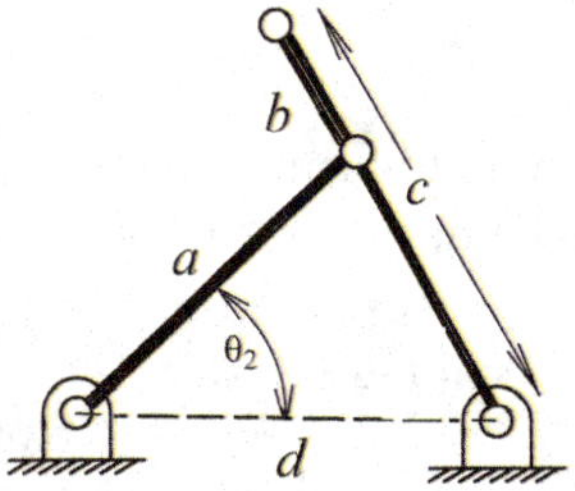	$\theta_2 = \cos^{-1}\left[\dfrac{d^2 + a^2 - (b-c)^2}{2ad}\right]$

Table 3.5: Toggle positions for non-Grashof double-rocker linkages

a = crank, b = coupler	c = rocker, d = ground
$a + b < c + d$ 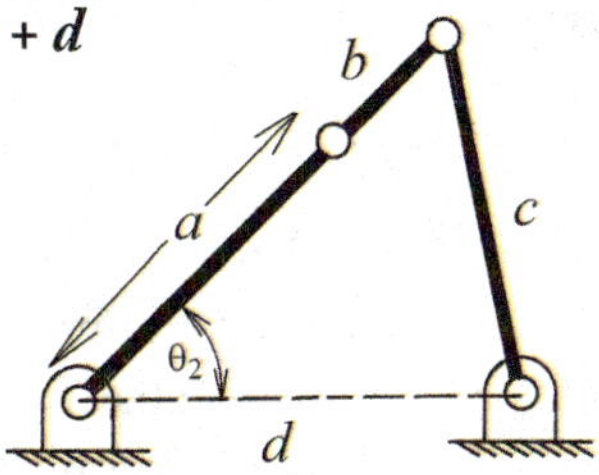	$\theta_2 = \cos^{-1}\left[\dfrac{(a + b)^2 + d^2 - c^2}{2d\,(a + b)}\right]$
or $a + b > c + d$ 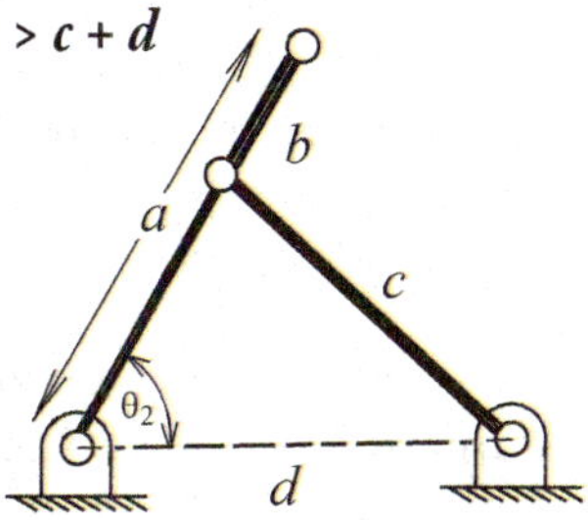	$\theta_2 = \cos^{-1}\left[\dfrac{(a - b)^2 + d^2 - c^2}{2d\,(a - b)}\right]$
$b + c < a + d$ 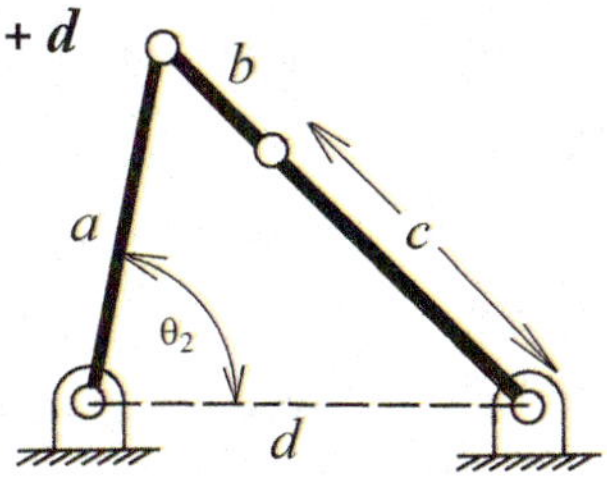	$\theta_2 = \cos^{-1}\left[\dfrac{d^2 + a^2 - (b + c)^2}{2ad}\right]$
or $b + c > a + d$ 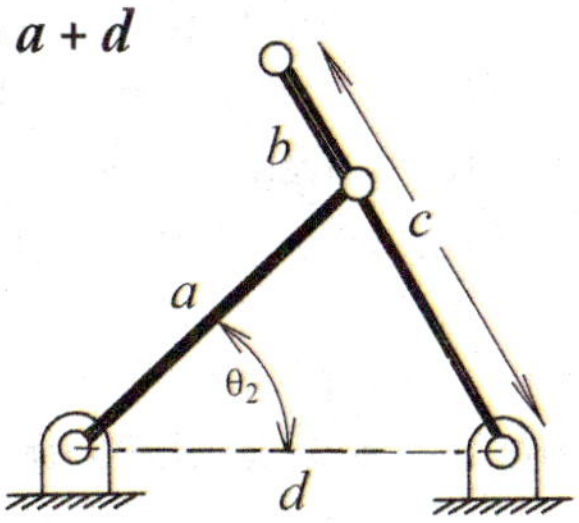	$\theta_2 = \cos^{-1}\left[\dfrac{d^2 + a^2 - (b - c)^2}{2ad}\right]$

(c) ***Toggle and mechanical advantage.*** Mechanical advantage is defined as:

$$\gamma = \frac{T_4}{T_2} = \frac{c \sin \mu}{a \sin \nu}.$$

(3.5)

From Equation (3.5) one observes that

when the angle μ approaches zero T_4 also approaches zero, regardless of T_2;

when the angle ν approaches zero T_4 approaches infinity, regardless of T_2.

(d) ***Applications of toggle positions.*** The observation of Equation (3.5) that when the angle ν approaches zero T_4 approaches infinity, regardless of T_2, has an important implication. The latter is that when the linkage is at or close to the toggle position its mechanical advantage is very large. Many mechanisms are designed and made based on the application of the concept of toggle position.

- Examples are: rock-crusher, vise grip locking pliers, and truck tailgate mechanism.

(e) ***Getting out of a toggle.***

- Grashof crank-rockers linkages: the crank will carry the linkage through the two toggle positions, and there is no need to perturb the rocker (because coupler and rocker will never assume co-linear positions).

- Others linkages: it is necessary to perturb one of the links that are in co-linear position.

3.6   CIRCUITS

Four-bar pin-jointed linkages have two alternate configurations for a given position of the input link; see Figures 3.7a and 3.7b.

The linkage cannot move from the first configuration to the second one without traveling through the toggle position when the coupler and the output link are co-linear; see Figure 3.7c.

If the toggle position can be negotiated, the two alternate configurations can be reached without taking the linkage apart (that is, physically disconnecting the joint between the coupler and the output link).

The *circuit* of a linkage is the number of possible configurations of motion that can only be obtained by physically disconnecting the joint between the coupler and the output link.

- All Grashof linkages have two circuits.

- All non-Grashof linkages have only one circuit.

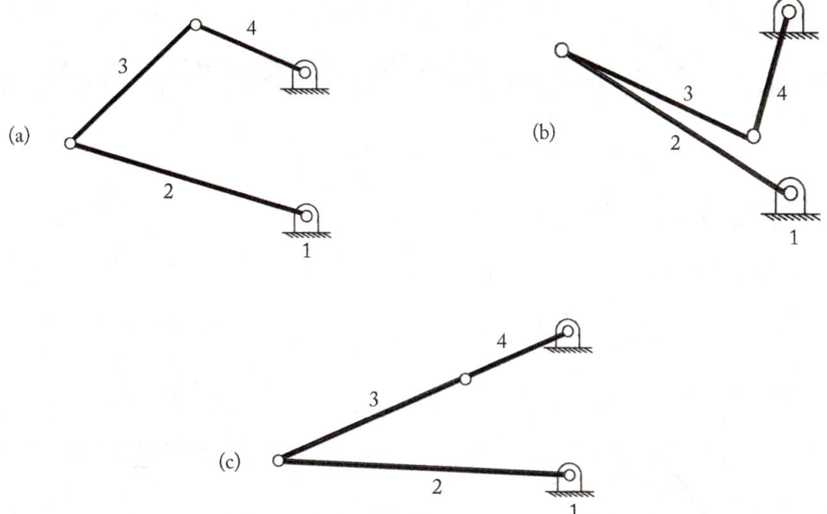

Figure 3.7: Configurations of linkages.

3.7   LIMITS OF ROTATION OF INPUT/OUTPUT LINKS

In the analysis and design of mechanisms it is important to know the limits of rotation of input/output links for many reasons. In particular, for safety consideration the limits of rotation are required so that the envelope of safety may be obtained.

In this section the limits of rotation of input/output links with reference to all types of four-bar mechanisms are presented in Sections 3.7.1–3.7.5.

3.7.1   GRASHOF CRANK-ROCKER LINKAGES

The input link, the crank, is able to make a full revolution. Thus,

$$0 \le \theta_2 \le 2\pi,$$

while the range of rotation of the output link, the rocker, is determined by the two toggle positions between the crank and coupler, as shown in Table 3.6.

3.7.2   GRASHOF DOUBLE-CRANK LINKAGES

The input and output links both are able to make full revolution. Thus,

$$0 \le \theta_2 \le 2\pi, \qquad 0 \le \theta_4 \le 2\pi.$$

Table 3.6: Limits of rotation of output link for Grashof crank-rocker linkages

a = crank, *b* = coupler	*c* = rocker, *d* = ground
	$\theta_4 = \pi - \cos^{-1}\left[\dfrac{d^2 + c^2 - (a+b)^2}{2cd}\right]$
	$\theta_4 = \pi - \cos^{-1}\left[\dfrac{d^2 + c^2 - (b-a)^2}{2cd}\right]$

3.7.3   GRASHOF ROCKER-CRANK LINKAGES

The rotation of link 2 is given by the relation

$$\beta \le \theta_2 \le \alpha,$$

where

$$\beta = \cos^{-1}\left[\frac{d^2 + a^2 - (b-c)^2}{2ad}\right] \quad \text{and} \quad \alpha = \cos^{-1}\left[\frac{d^2 + a^2 - (b+c)^2}{2ad}\right]$$

while the range of rotation of link 4, the crank is

$$0 \le \theta_4 \le 2\pi.$$

That is, α is concerned with extended co-linearity between links 3 and 4 while β has to do with overlapping co-linearity between links 3 and 4.

An example is given in Figure 3.8.

3.7.4   GRASHOF DOUBLE-ROCKER LINKAGES

The rotation of link 2 is given by the relation

$$\beta \le \theta_2 \le \alpha,$$

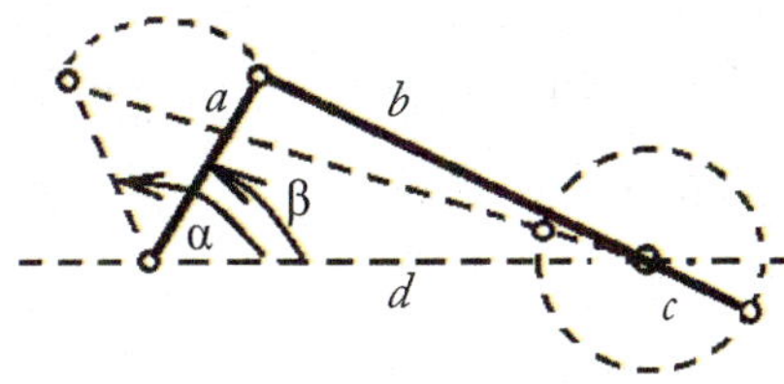

Figure 3.8: Definitions of α and β in a Grashof rocker-crank linkage: $a = 4$, $b = 11$, $c = 2$, and $d = 10$.

where α and β are defined as those in the Grashoff rocker-crank case while the range of rotation of link 4, is given by

$$\pi - A \leq \theta_4 \leq \pi - B$$

in which $A = \cos^{-1}\left[\frac{d^2+c^2-(a+b)^2}{2cd}\right]$ and $B = \cos^{-1}\left[\frac{d^2+c^2-(a-b)^2}{2cd}\right]$.

That is, A is concerned with extended co-linearity between links 2 and 3 while B has to do with overlapping co-linearity between links 2 and 3.

An example is presented in Figure 3.9.

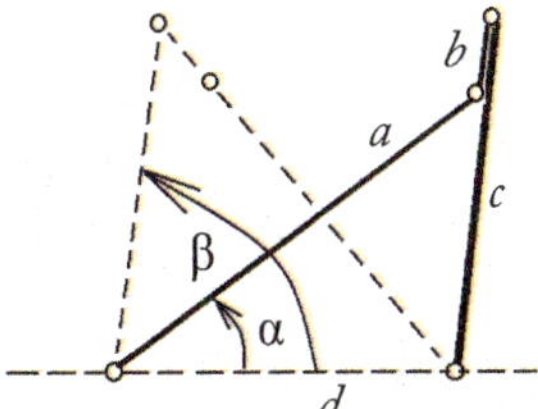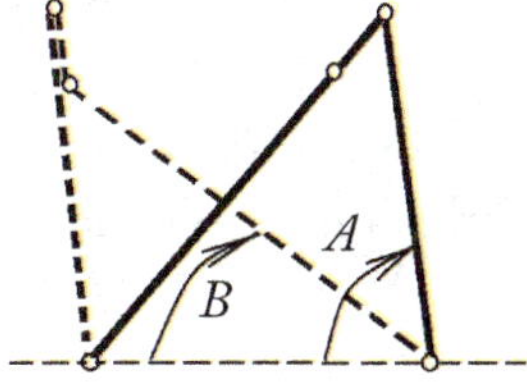

Figure 3.9: Definitions of α, β, A, and B in a Grashof double-rocker linkage: $a = 53$, $b = 15$, $c = 48$, and $d = 40$.

3.7.5   NON-GRASHOF DOUBLE-ROCKER LINKAGES

In this type of linkages the limits of rotation of input/output links have two cases.

Case (a) Non-Grashof double-rocker linkages.   In this case links 3 and 4 are in extended co-linearity with either one of the following two conditions is satisfied: $d + a > b + c$ and $|d - a| > |b - c|$ then

$$-\alpha \leq \theta_2 \leq \alpha,$$

where

$$\alpha = \cos^{-1}\left[\frac{a^2 + d^2 - (c + b)^2}{2ad}\right]$$

and

$$\pi - A \le \theta_4 \le \pi + A \quad \text{or} \quad \pi - B \le \theta_4 \le \pi + B$$

in which A and B are defined as in Section 3.7.4, and use B when $a + b > c + d$ is true or use A when $a + b < c + d$ is true.

Example 3.3

When $a = 4, b = 8, c = 5, d = 10$ the angles are illustrated in Figure 3.10.

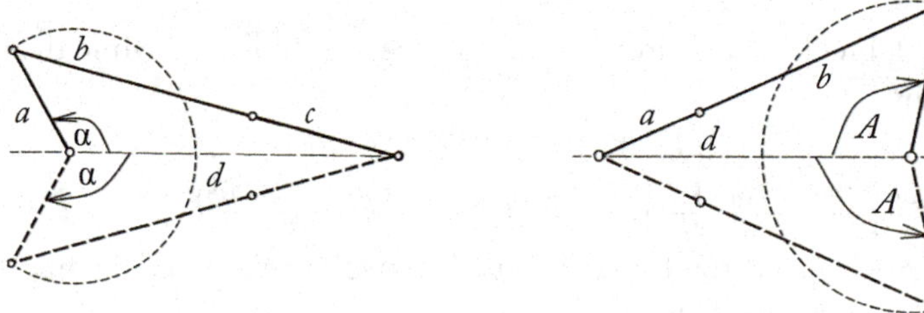

Figure 3.10: Definitions of angles α and A.

Case (b) Non-Grashof double-rocker linkages.   In this case, links 3 and 4 are in overlapping co-linearity with either one of the following two conditions is satisfied: $d + a < b + c$ and $|d - a| < |b - c|$ then

$$-\beta \le \theta_2 \le 2\pi - \beta$$

where

$$\beta = \cos^{-1}\left[\frac{a^2 + d^2 - (b - c)^2}{2ad}\right]$$

and $\pi - A \le \theta_4 \le \pi + A$ or

$$\pi - B \le \theta_4 \le \pi + B$$

in which A and B are defined as in Section 3.7.4.

Example 3.4

When $a = 8, b = 5, c = 10, d = 4$ the angles are illustrated in Figure 3.11.

3.8   SUMMARY

The results in the foregoing sections are summarized in Table 3.7. The angles α, β, A, and B for θ_2 and θ_4 are those defined in Section 3.7.

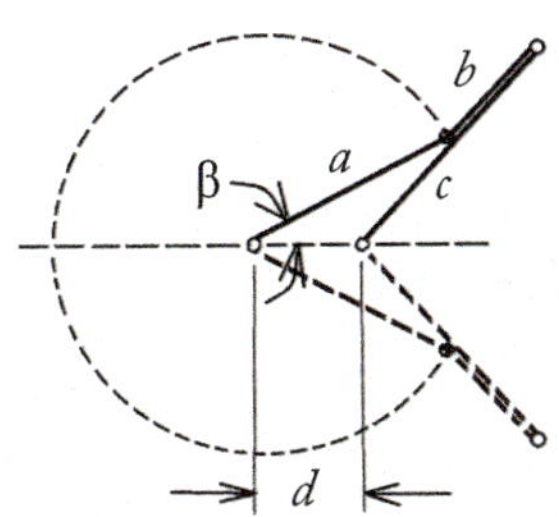 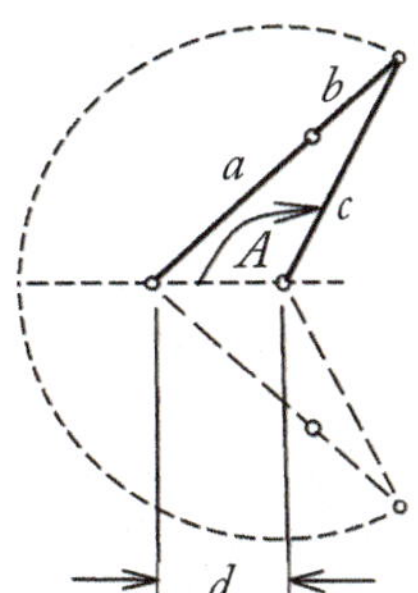

Figure 3.11: Definitions of angles β and A.

Table 3.7: Summary of four-bar pin-jointed linkages

Existance Condition	$L \leq S + P + Q$				
Grashof Condition	$S + L \leq P + Q$				$S + L > P + Q$
Type	Grashof crank-rocker	Grashof rocker-crank	Grashof double-crank	Grashof double-rocker	Non-Grashof double-rocker
Grounded link	Either one adjacent "S"		"S"	The link opposite "S"	Any one
μ	$\mu_{min} \sim \mu_{max}$	$0° \sim 90°$	$0° \sim 90°$	$0° \sim 90°$	$0° \sim \mu_{max}$
Toggles	2	2	0	4	2
θ_2	$0° \sim 360°$	$\beta \sim \alpha$	$0° \sim 360°$	$\beta \sim \alpha$	$-\alpha \sim \alpha$ $180°-\beta \sim 180°+\beta$
θ_4	$180°-A \sim 180°-B$	$0° \sim 360°$	$0° \sim 360°$	$180°-A \sim 180°-B$	$180°-A \sim 180°+A$ $180°-B \sim 180°+B$

CHAPTER 4

Analysis of Planar Mechanisms

In the last chapter the focus was on the analysis of four-bar mechanisms. It should be pointed out that there are many linkages having more than four-bars. Typical examples are the *Watt's six-bar* and *Stephenson's six-bar* mechanisms. The former and latter mechanisms are shown in Figure 4.1. As mentioned in the last chapter, the concepts for four-bar mechanisms can be applied to the analysis of linkages having more than four-bars.

(a) (b)

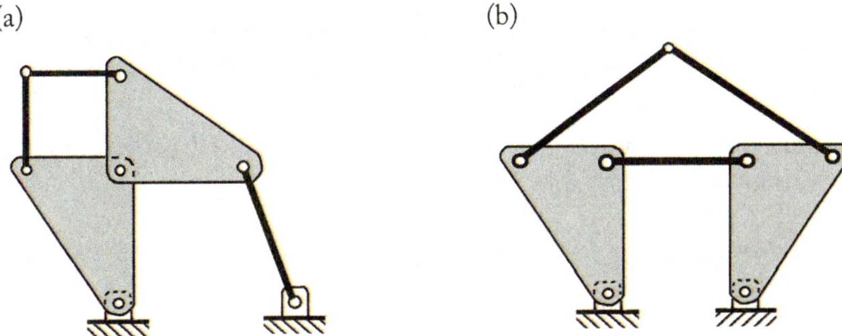

Figure 4.1: Six-bar mechanisms: (a) Watt's six-bar linkage and (b) Stephenson's six-bar linkage.

In order to demonstrate the applications of the concepts introduced in the last chapter one planar four-bar mechanism is presented in Section 4.1 while Section 4.2 deals with a planar six-bar mechanism.

4.1 ANALYSIS OF FOUR-BAR MECHANISMS

In this section an example of four-bar mechanism is included to illustrate the use of the concepts presented in Chapter 3.

Example 4.1
The lengths of a four-bar linkage are: $L_1 = d = 52.0$ units, $L_2 = a = 8.0$ units, $L_3 = b = 50.0$ units, and $L_4 = c = 25.0$ units.

(a) Determine the type of the linkage.

(b) Determine the range of transmission angle.

(c) Determine the toggle position, in term of θ_2.

(d) Identify the range of rotation of link 2.

(e) Determine, graphically, the range of rotation of link 4.

Solution:

(a) *Type of linkage*

$S = L_2 = 8.0$ units, $L = L_1 = 52.0$ units. Therefore,

$$S + L = 60.0 \text{ units},$$

$$P + Q = L_3 + L_4 = 50.0 + 25.0 \text{ units} = 75.0 \text{ units}.$$

That is, $S + L < P + Q$ and the linkage is Grashoff.

Since the shortest link, L_2, is adjacent to the ground link, L_1, therefore the linkage is a Grashof crank-rocker (see Table 3.7).

(b) *Transmission angle*

$a = 8.0, b = 50.0, c = 25.0, d = 52.0$ and according to Table 3.1,

$$\mu_1 = \cos^{-1}\left[\frac{b^2 + c^2 - (d + a)^2}{2bc}\right]$$

$$= \cos^{-1}\left[\frac{50.0^2 + 25.0^2 - (52.0 + 8.0)^2}{2(50.0)(25.0)}\right]$$

$$= \cos^{-1}(-0.1900) = 100.95°.$$

Using acute angle, hence $\mu_1 = 180° - 100.95° = 79.05°$.

$$\mu_2 = \cos^{-1}\left[\frac{b^2 + c^2 - (d - a)^2}{2bc}\right]$$

$$= \cos^{-1}\left[\frac{50.0^2 + 25.0^2 - (52.0 - 8.0)^2}{2(50.0)(25.0)}\right]$$

$$= 61.60°.$$

Therefore,

$$\mu_{\min} = \mu_2 = 61.60°, \qquad \mu_{\max} = \mu_1 = 79.05°.$$

Note that $\mu_{\min} > 35°$.

(c) *Toggle positions*

Since the present linkage is a Grashof crank-rocker and therefore from Table 3.3,

$$\theta_2' = \theta_2 = \cos^{-1}\left[\frac{d^2 + (a+b)^2 - c^2}{2d(a+b)}\right]$$

$$= \cos^{-1}\left[\frac{52.0^2 + (8.0 + 50.0)^2 - 25.0^2}{2(52.0)(8.0 + 50.0)}\right]$$

$$= \cos^{-1}(0.9024) = 25.52°.$$

The other toggle position is defined by

$$\theta_2'' = \theta_2 = -\cos^{-1}\left[\frac{d^2 + (a-b)^2 - c^2}{2d(a-b)}\right]$$

$$= -\cos^{-1}\left[\frac{52.0^2 + (8.0 - 50.0)^2 - 25.0^2}{2(52.0)(8.0 - 50.0)}\right]$$

$$= -\cos^{-1}(-0.8798) = -151.62°.$$

(d) *Range of rotation of link 2*

Since it is a Grashof crank-rocker and therefore the range of rotation of link 2 is

$$0 \le \theta_2 \le 2\pi.$$

(e) *Range of rotation of link 4*

Graphically, see Figure 4.2 in which a scale has been chosen for the construction,

$$\theta_2' \quad \text{associates with} \quad \theta_4' = 89.0°$$
$$\theta_2'' \quad \text{associates with} \quad \theta_4'' = 127.0°.$$

Therefore, the range of rotation of link 4 is

$$89.0° \le \theta_4 \le 127.0°.$$

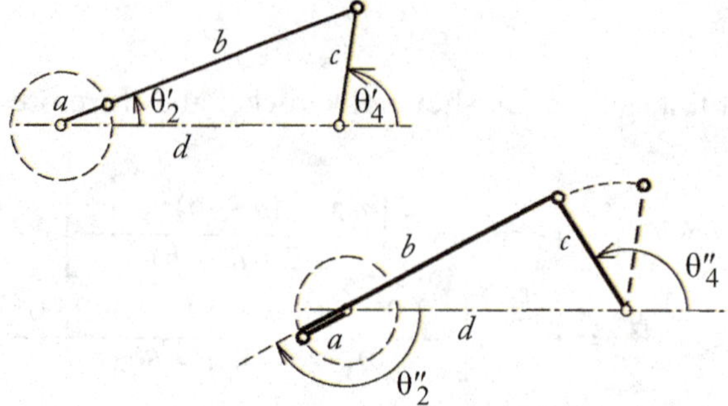

Figure 4.2: Range of rotation of link 4.

Analytically,

$$\theta_4' = 180° - \cos^{-1}\left[\frac{d^2 + c^2 - (a+b)^2}{2cd}\right]$$

$$= 180° - \cos^{-1}\left[\frac{52.0^2 + 25.0^2 - (8.0 + 50.0)^2}{2(25.0)(52.0)}\right]$$

$$\theta_4' = 180° - \cos^{-1}[-0.0135] = 180° - 90.77° = 89.23°.$$

$$\theta_4'' = 180° - \cos^{-1}\left[\frac{d^2 + c^2 - (b-a)^2}{2cd}\right]$$

$$= 180° - \cos^{-1}\left[\frac{52.0^2 + 25.0^2 - (50.0 - 8.0)^2}{2(52.0)(25.0)}\right]$$

$$= 180° - \cos^{-1}(0.6019) = 180° - 52.99° = 127.01°.$$

4.2   ANALYSIS OF SIX-BAR MECHANISMS

This section is concerned the application of the concepts for four-bar linkages to linkages that have more than four bars. For brevity, one six-bar linkage is considered and studied in the following.

Example 4.2

Consider the six-bar pin-jointed linkage shown in Figure 4.3. This linkage can be considered as consisting of two sub-four-bar planar linkages, O_2ABO_4 and O_4CDO_6. Note that the links are not drawn to scale in the figure. The lengths of the links for the first four-bar linkage are: $L_{1a} = O_2O_4 = 3.98$, $L_2 = O_2A = 0.69$, $L_3 = AB = 3.63$, and $L_{4a} = BO_4 = 1.82$. The lengths of the links of the second sub-four-bar planar linkage are: $L_{1b} = O_4O_6 = 6.23$, $L_{4b} = O_4C = 3.56$, $L_5 = CD = 2.48$, and $L_6 = DO_6 = 4.30$. For each of the sub-four-bar linkage,

(a) check if it is Grashof;

(b) identify the type of linkage;

(c) determine the toggle positions and show the results graphically (by the angle between respective ground link and input link); and

(d) evaluate the maximum transmission angle and draw the corresponding configuration.

Solution:

(a) *Grashof condition*

For sub-four-bar linkage O_2ABO_4, $a = 0.69$, $b = 3.63$, $c = 1.82$, and $d = 3.98$, therefore, $S = 0.69$, $L = 3.98$, $P = 1.82$, and $Q = 3.63$,

$$S + L = 4.67, \quad P + Q = 5.45, \quad S + L < P + Q.$$

Therefore, this linkage is Grashof.

For sub-four-bar linkage, O_4CDO_6, $a = 3.56$, $b = 2, 48$, $c = 4.30$, and $d = 6.23$, therefore, $S = 2.48$, $L = 6.23$, $P = 3.56$, and $Q = 4.30$,

$$S + L = 8.71, \quad P + Q = 7.86, \quad S + L > P + Q.$$

Therefore, this linkage is non-Grashof.

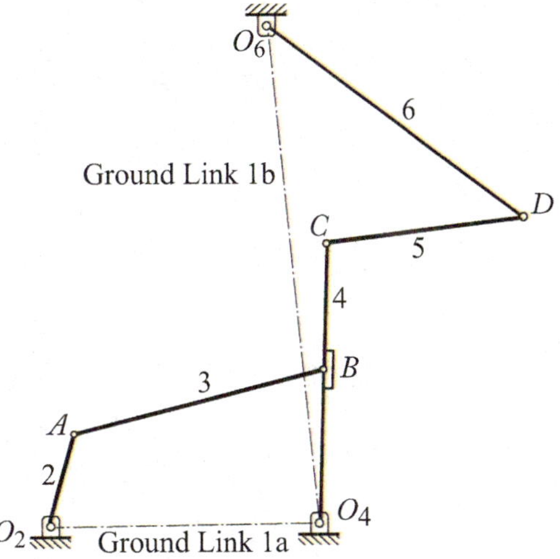

Figure 4.3: Six-bar mechanism.

(b) *Type of linkage*

For sub-four-bar linkage O_2ABO_4, since the longest link is grounded and is adjacent to the shortest link it is a Grashof crank-rocker (see Table 3.7). Link 2 is the crank.

For sub-four-bar linkage O_4CDO_6, it is a non-Grashof double-rocker (see Table 3.7). Note that all inversion of a non-Grashof four-bar linkage is double-rocker.

(c) *Toggle positions*

For sub-four-bar linkage O_2ABO_4, since it is a Grashof crank-rocker and therefore from Table 3.3,

$$\theta_2' = \theta_2 = \cos^{-1}\left[\frac{d^2 + (a+b)^2 - c^2}{2d(a+b)}\right]$$

$$\theta_2' = \cos^{-1}\left[\frac{3.98^2 + (0.69 + 3.63)^2 - 1.82^2}{2(3.98)(0.69 + 3.63)}\right]$$

$$= \cos^{-1}(0.907) = 24.90°.$$

The other toggle position is defined by

$$\theta_2'' = \theta_2 = -\cos^{-1}\left[\frac{d^2 + (a-b)^2 - c^2}{2d(a-b)}\right]$$

$$= -\cos^{-1}\left[\frac{3.98^2 + (0.69 - 3.63)^2 - 1.82^2}{2(3.98)(0.69 - 3.63)}\right]$$

$$= -\cos^{-1}(-0.905) = -154.80°.$$

Graphically, by using a scale, the two toggle positions are shown in Figure 4.4.

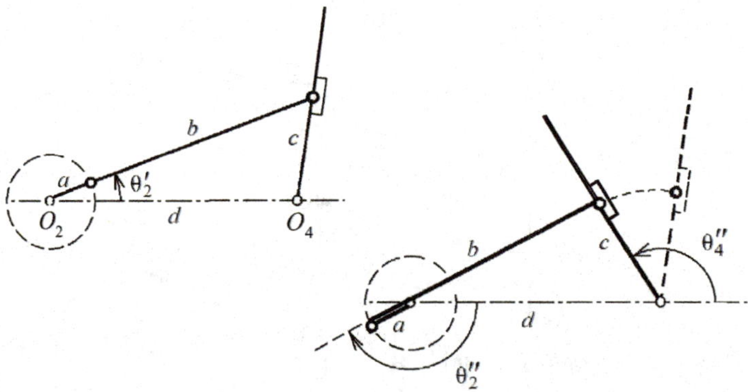

Figure 4.4: Toggle positions of first sub-four-bar linkage.

For sub-four-bar linkage O_4CDO_6, since it is a non-Grashof double-rocker and hence, from Tables 3.5 and 3.7, one has, graphically (by using a scale), the two toggle positions are shown in Figure 4.5.

Analytically, when links 4 and 5 are in extended co-linearity

$$\theta_2' = \theta_2 = \cos^{-1}\left[\frac{d^2 + (a+b)^2 - c^2}{2d(a+b)}\right]$$

$$= \cos^{-1}\left[\frac{6.23^2 + (3.56 + 2.48)^2 - 4.30^2}{2(6.23)(3.56 + 2.48)}\right]$$

$$= \cos^{-1}(0.755) = 41.00°.$$

The other toggle position is defined by the extended co-linearity of links 5 and 6,

$$\theta_2'' = \theta_2 = \cos^{-1}\left[\frac{a^2 + d^2 - (b+c)^2}{2ad}\right]$$

$$= \cos^{-1}\left[\frac{3.56^2 + 6.23^2 - (2.48 + 4.20)^2}{2(3.56)(6.23)}\right]$$

$$= cos^{-1}(0.124) = 82.90°.$$

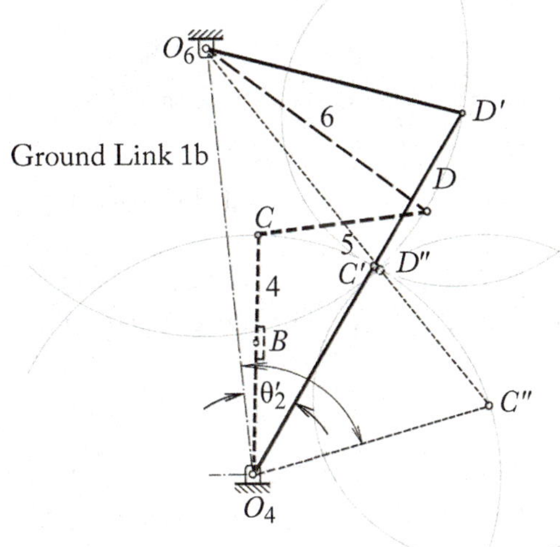

Figure 4.5: Toggle positions of second sub-four-bar linkage.

(d) *Transmission angle*

For sub-four-bar $O_2ABO_4, a = 0.69, b = 3.63, c = 1.82, d = 3.98$ and according to Table 3.1,

$$\mu_1 = \cos^{-1}\left[\frac{b^2 + c^2 - (d+a)^2}{2bc}\right]$$

$$= \cos^{-1}\left[\frac{3.63^2 + 1.82^2 - (3.98 + 0.69)^2}{2(3.63)(1.82)}\right]$$

$$= \cos^{-1}[-0.403] = 113.70°.$$

Using acute angle, hence $\mu_1 = 180° - 113.70° = 66.3°$.

$$\mu_2 = \cos^{-1}\left[\frac{b^2 + c^2 - (d-a)^2}{2bc}\right]$$

$$= \cos^{-1}\left[\frac{3.63^2 + 1.82^2 - (3.98 - 0.69)^2}{2(3.63)(1.82)}\right]$$

$$= \cos^{-1}(0.429) = 64.60°.$$

Therefore, $\mu_{\max} = \mu_1 = 66.30°$ which is obtained when links 1a and 2 are in extended co-linearity. This is sketched in Figure 4.6 by using a scale for the construction.

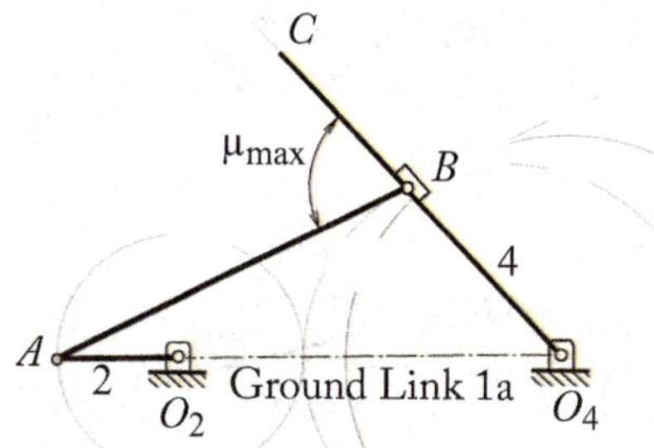

Figure 4.6: Maximum transmission angle of first sub-four-bar linkage.

For sub-four-bar linkage O_4CDO_6, since it is a non-Grashof double-rocker and $a + b < c + d$ therefore from Table 3.2 $\mu_{\max}$ is obtained when links 4 and 5 are in extended co-linearity. That is,

$$\mu_{\max} = \cos^{-1}\left[\frac{(a+b)^2 + c^2 - d^2}{2c(a+b)}\right]$$

$$= \cos^{-1}\left[\frac{(3.56 + 2.48)^2 + 4.30^2 - 6.23^2}{2(4.30)(3.56 + 2.48)}\right]$$

$$= \cos^{-1}(0.311) = 71.90°.$$

The scaled construction for this case is given in Figure 4.7.

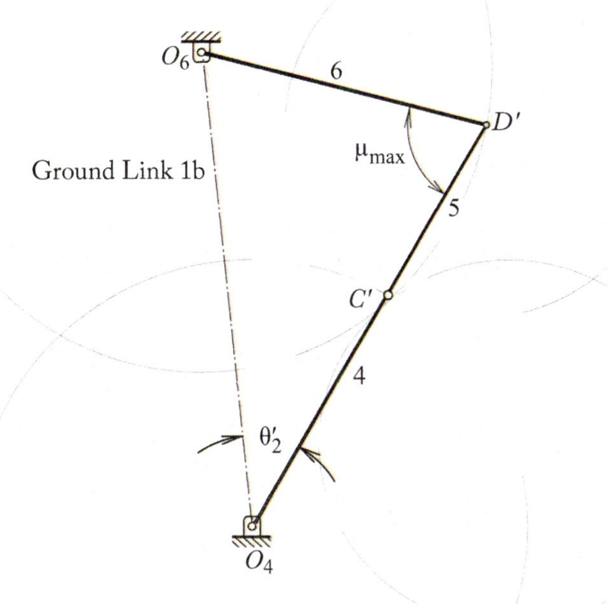

Figure 4.7: Maximum transmission angle of second sub-four-bar linkage.

4.3   EXERCISES

4.1.   The lengths of a four-bar linkage are: $L_1 = d = 22.0$ units, $L_2 = a = 34.0$ units, $L_3 = b = 19.0$ units, and $L_4 = c = 21.0$ units.

   (a) Determine the type of the linkage.

   (b) Determine the range of transmission angle.

   (c) Find the limits of rotation of link 2.

   (d) Find the limits of rotation of link 4.

4.2.   The lengths of a four-bar linkage are: $L_1 = d = 53.0$ units, $L_2 = a = 9.50$ units, $L_3 = b = 48.0$ units, and $L_4 = c = 25.50$ units.

   (a) Determine the type of the linkage.

   (b) Evaluate the range of transmission angle.

   (c) Find the limits of rotation of link 2.

   (d) Find the limits of rotation of link 4.

4.3. The lengths of a four-bar linkage are: $L_1 = d = 23.0$ units, $L_2 = a = 33.50$ units, $L_3 = b = 20.0$ units, and $L_4 = c = 20.50$ units.

    (a) Determine the type of the linkage.

    (b) Determine the range of transmission angle.

    (c) Find the limits of rotation of link 2.

    (d) Find the limits of rotation of link 4.

4.4. A quick-return mechanism is shown in Figure 4.8. Derive an expression for the displacement y of the slider (link 5) as a function only of θ of the input link (link 2) and the constant horizontal distances x_1 and x_2.

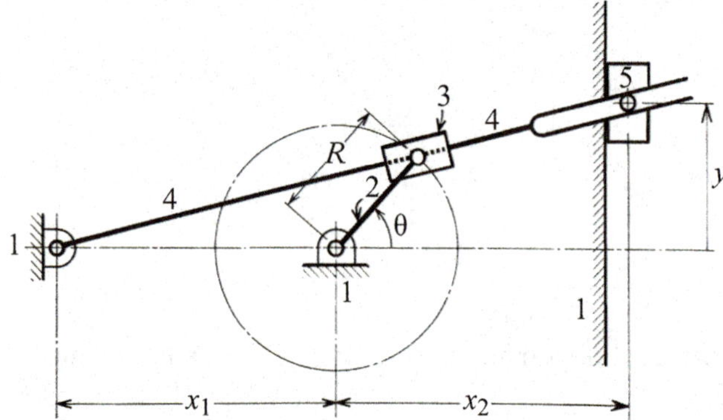

Figure 4.8: A quick-return mechanism.

CHAPTER 5

Analytical Approaches in Motion Studies

There are various analytical approaches in the studies of motion in machinery. However, in this chapter only two of them are presented. They are the method of complex number and the unit vector or rotating frame of reference (RFR) solution method. These two approaches are included in Sections 5.1–5.3 while Section 5.4 is concerned with representative questions and solutions.

5.1 METHOD OF COMPLEX NUMBER

This method is straightforward and can be applied to the motion analysis of spatial mechanisms. However, in the present chapter it is applied to the motion analysis of planar mechanisms or mechanisms in two-dimensional (2D) space.

In the following two sections various common representations of a vector in 2D space and the detail steps of the method are presented.

5.1.1 REPRESENTATIONS OF A PLANAR VECTOR

Consider the planar vector in Figure 5.1. Various common representations exist in the literature. They are the polar, Cartesian, and complex number form representations. Their formulas are given below.

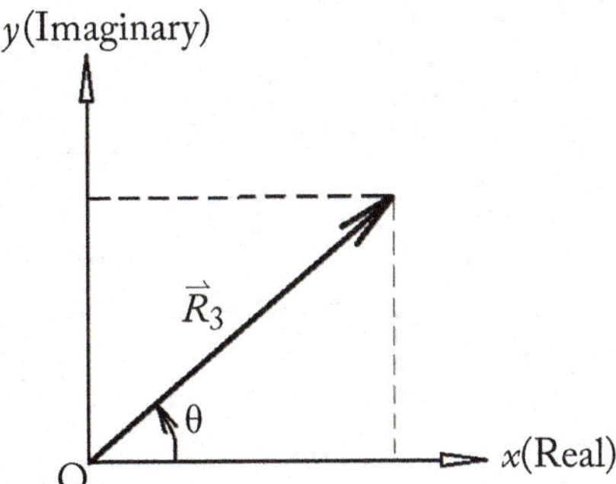

Figure 5.1: Vector representation in complex number form.

In *Polar form*:

$$\vec{R} = r \,@\, \angle\theta,$$

where r is the magnitude, and θ is the phase of the vector.

In *Cartesian form*:

$$\vec{R} = r\cos\theta\,\vec{\imath} + r\sin\theta\,\vec{\jmath},$$

where $\vec{\imath}$ and $\vec{\jmath}$ are the unit vectors along the x and y axes of the 2D Cartesian co-ordinate system.

In *Complex number form*:

$$\vec{R} = re^{i\theta} = r\cos\theta + ir\sin\theta$$

in which i is the imaginary number. That is, $i = \sqrt{-1}$ and should not be confused with the unit vector $\vec{\imath}$. The expression $e^{i\theta} = \cos\theta + i\sin\theta$ is known as *Euler's identity*.

Euler's identities:

$$\pm e^{i\theta} = \cos\theta \pm i\sin\theta,$$
$$\pm i\,e^{i\theta} = \mp\sin\theta \pm i\cos\theta.$$

Multiplication of a vector by i:

$$
\begin{aligned}
i\vec{R} = ire^{i\theta} &= ir(\cos\theta + i\sin\theta) = r(-\sin\theta + i\cos\theta) \\
&= r\left[\cos\left(90° + \theta\right) + i\sin\left(90° + \theta\right)\right] \\
&= re^{i(90° + \theta)}.
\end{aligned}
$$

This means that by multiplying the vector $\vec{R}$ with the imaginary number i it is equivalent to rotating the vector by 90° counterclockwise (ccw), or $i\vec{R}$ represents a vector that has the same magnitude as $\vec{R}$ and is normal to $\vec{R}$.

The following slider crank mechanism is included to illustrate the simple steps in the complex number method and the use of Euler's identities.

5.1.2   EXAMPLE

The slider crank mechanism shown in Figure 5.2 has the following data: the length of the input link 2 $= a$, link 3 $= b$, link 1 $= d$; the given angular position, velocity, and acceleration of link 2 are, respectively, θ_2, ω_2, and α_2. Determine all the unknowns in this mechanism.

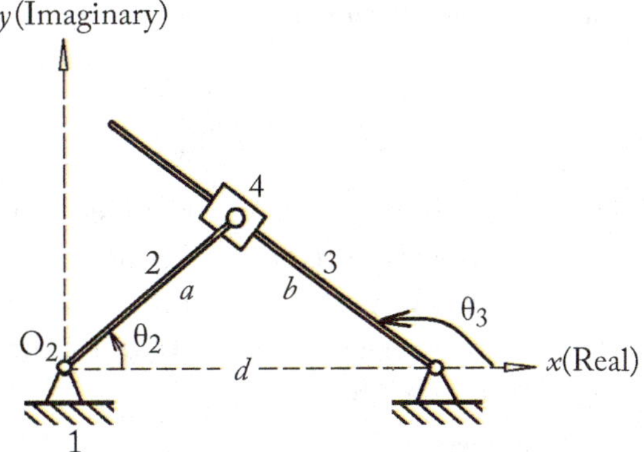

Figure 5.2: Slider crank mechanism.

Solution:
The solution consists of three stages; the position solution, velocity solution, and acceleration solution.

(a) **Position solution**

The vector loop diagram for the above mechanism is shown in Figure 5.3.

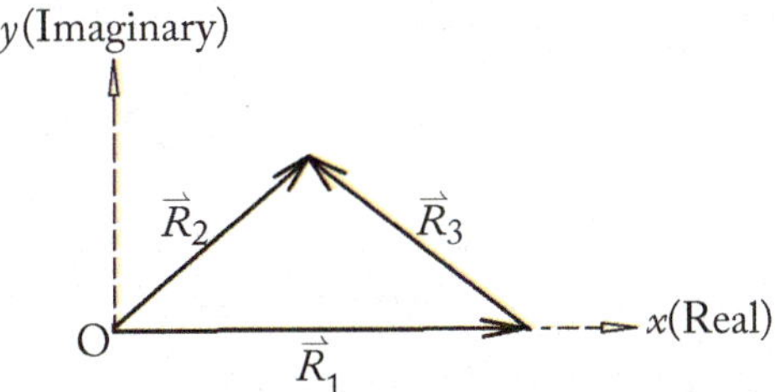

Figure 5.3: Vector loop diagram.

The vector loop equation from the vector loop diagram is given by

$$\vec{R}_1 - \vec{R}_2 + \vec{R}_3 = \vec{0}. \tag{5.1}$$

It should be noted that if the direction of a particular vector is incorrectly assumed at this stage one can still proceed with the analysis. At the end of the analysis one would discover that the direction is negative which, in turn, means that the direction of that particular vector in Equation (5.1) is just opposite to the true one.

Complex number representation of Equation (5.1) gives

$$de^{i\theta_1} - ae^{i\theta_2} + be^{i\theta_3} = 0. \tag{5.2}$$

Application of Euler identity, also noting $\theta_1 = 0$, Equation (5.2) becomes

$$d - a\cos\theta_2 - ia\sin\theta_2$$
$$+ b\cos\theta_3 + ib\sin\theta_3 = 0. \tag{5.3}$$

Note that, aside from the first term on the left-hand side (lhs) of Equation (5.2), the second and third terms have been expanded in one term per line in Equation (5.3). This is because in doing so one can easily inspect whether or not there is an error in the operation of the equation. This way of expanding one term per line for the complex number method will be followed throughout this chapter.

Separation of real and imaginary parts,

$$d - a\cos\theta_2 + b\cos\theta_3 = 0 \tag{5.4}$$
$$-a\sin\theta_2 + b\sin\theta_3 = 0. \tag{5.5}$$

The unknowns are b and θ_3.

Solving for b by rewriting Equations (5.4) and (5.5) as

$$d - a\cos\theta_2 = -b\cos\theta_3 \tag{5.6}$$
$$-a\sin\theta_2 = -b\sin\theta_3. \tag{5.7}$$

Taking the square of both sides of Equations (5.6) and (5.7), and adding the resulting equations, one has

$$(d - a\cos\theta_2)^2 + (a\sin\theta_2)^2 = b^2.$$

Therefore,

$$b = \sqrt{(d - a\cos\theta_2)^2 + (a\sin\theta_2)^2}. \tag{5.8}$$

Solving for θ_3 by dividing Equation (5.7) by (5.6),

$$\frac{-a\sin\theta_2}{d - a\cos\theta_2} = \frac{\sin\theta_3}{\cos\theta_3} = \tan\theta_3. \tag{5.9}$$

Equations (5.8) and (5.9) are the position solutions.

(b) *Velocity solution*

Differentiation of Equation (5.2) with respect to time t, gives

$$-a\left(i\dot{\theta}_2\right)e^{i\theta_2} + \dot{b}e^{i\theta_3} + b\left(i\dot{\theta}_3\right)e^{i\theta_3} = 0. \tag{5.10}$$

Application of Euler identity, and noting that the over-dot denotes derivative with respect to time t, such as

$$\dot{\theta}_2 = \frac{d\theta_2}{dt} = \omega_2, \quad \dot{\theta}_3 = \frac{d\theta_3}{dt} = \omega_3.$$

Equation (5.10) becomes

$$-a\omega_2\left(-\sin\theta_2 + i\cos\theta_2\right)$$
$$+\dot{b}\left(\cos\theta_3 + i\sin\theta_3\right)$$
$$+b\omega_3\left(-\sin\theta_3 + i\cos\theta_3\right) = 0.$$

Separation of real and imaginary parts,

$$-a\omega_2\left(-\sin\theta_2\right) + \dot{b}\left(\cos\theta_3\right) + b\omega_3\left(-\sin\theta_3\right) = 0, \tag{5.11}$$

$$-a\omega_2\left(\cos\theta_2\right) + \dot{b}\left(\sin\theta_3\right) + b\omega_3\left(\cos\theta_3\right) = 0. \tag{5.12}$$

The unknowns are $\dot{b}$ and ω_3.

Solving for $\dot{b}$ by performing (5.11) $\cos\theta_3$ + (5.12) $\sin\theta_3$ so that

$$a\omega_2\left(\sin\theta_2\cos\theta_3 - \cos\theta_2\sin\theta_3\right) + \dot{b} = 0.$$

Therefore,

$$\dot{b} = -a\omega_2\sin\left(\theta_2 - \theta_3\right). \tag{5.13}$$

Solving for ω_3 by performing (5.11) $\sin\theta_3-$ (5.12) $\cos\theta_3$ so that

$$a\omega_2\left(\sin\theta_2\sin\theta_3 + \cos\theta_2\cos\theta_3\right) - b\omega_3 = 0.$$

Therefore,

$$\omega_3 = \left(\frac{a\omega_2}{b}\right)\cos\left(\theta_2 - \theta_3\right). \tag{5.14}$$

Equations (5.13) and (5.14) are the velocity solutions.

(c) *Acceleration solution*

Differentiation of Equation (5.10) with respect to time t, one obtains

$$-a\left(i\ddot{\theta}_2\right)e^{i\theta_2} - a\left(i\dot{\theta}_2\right)^2 e^{i\theta_2}$$
$$+\ddot{b}e^{i\theta_3} + \dot{b}\left(i\dot{\theta}_3\right)e^{i\theta_3}$$
$$+\dot{b}\left(i\dot{\theta}_3\right)e^{i\theta_3} + b\left(i\dot{\theta}_3\right)^2 e^{i\theta_3} + b\left(i\ddot{\theta}_3\right)e^{i\theta_3} = 0.$$

Writing $\frac{d^2\theta_2}{dt^2} = \alpha_2$ and $\frac{d^2\theta_3}{dt^2} = \alpha_3$, it leads to

$$-a\left(i\alpha_2\right)e^{i\theta_2} + a\left(\omega_2\right)^2 e^{i\theta_2}$$
$$+\ddot{b}e^{i\theta_3} + 2\dot{b}\left(i\omega_3\right)e^{i\theta_3}$$
$$-b\left(\omega_3\right)^2 e^{i\theta_3} + b\left(i\alpha_3\right)e^{i\theta_3} = 0.$$

Application of Euler identity,

$$- a\alpha_2 \left(- \sin \theta_2 + i \cos \theta_2\right)$$

$$+ a \left(\omega_2\right)^2 \left(\cos \theta_2 + i \sin \theta_2\right)$$

$$+ \left(\ddot{b} - b\omega_3^2\right) \left(\cos \theta_3 + i \sin \theta_3\right)$$

$$+ \left(b\alpha_3 + 2\dot{b}\omega_3\right) \left(- \sin \theta_3 + i \cos \theta_3\right) = 0.$$

Separation of real and imaginary parts,

$$a\alpha_2 \left(\sin \theta_2\right) + a \left(\omega_2\right)^2 \left(\cos \theta_2\right)$$

$$+ \left(\ddot{b} - b\omega_3^2\right) \left(\cos \theta_3\right)$$

$$- \left(b\alpha_3 + 2\dot{b}\omega_3\right) \left(\sin \theta_3\right) = 0, \tag{5.15}$$

$$- a\alpha_2 \left(\cos \theta_2\right) + a \left(\omega_2\right)^2 \left(\sin \theta_2\right)$$

$$+ \left(\ddot{b} - b\omega_3^2\right) \left(\sin \theta_3\right)$$

$$+ \left(b\alpha_3 + 2\dot{b}\omega_3\right) \left(\cos \theta_3\right) = 0. \tag{5.16}$$

The unknowns are $\ddot{b}$ and α_3.

Solving for $\ddot{b}$ by performing (5.15) $\cos \theta_3 +$ (5.16) $\sin \theta_3$ so that

$$a\alpha_2 \left(\sin \theta_2 \cos \theta_3 - \cos \theta_2 \sin \theta_3\right)$$

$$+ a\omega_2^2 \left(\sin \theta_2 \sin \theta_3 + \cos \theta_2 \cos \theta_3\right)$$

$$+ \left(\ddot{b} - b\omega_3^2\right) = 0.$$

Therefore,

$$\ddot{b} = -a\alpha_2 \sin \left(\theta_2 - \theta_3\right) - a\omega_2^2 \cos \left(\theta_2 - \theta_3\right) + b\omega_3^2. \tag{5.17}$$

Solving for α_3 by performing (5.15) $\sin \theta_3 -$ (5.16) $\cos \theta_3$ so that

$$- a\alpha_2 \left(\sin \theta_2 \sin \theta_3 + \cos \theta_2 \cos \theta_3\right)$$

$$+ a\omega_2^2 \left(\cos \theta_2 \sin \theta_3 - \sin \theta_2 \cos \theta_3\right)$$

$$+ \left(b\alpha_3 + 2\dot{b}\omega_3\right) = 0.$$

Therefore,

$$\alpha_3 = \left(\frac{a\alpha_2}{b}\right) \cos \left(\theta_2 - \theta_3\right) - \left(\frac{a\omega_2^2}{b}\right) \sin \left(\theta_2 - \theta_3\right) - \left(\frac{2\dot{b}\omega_3}{b}\right). \tag{5.18}$$

Equations (5.17) and (5.18) are the acceleration solutions.

5.2   SUMMARY OF COMPLEX NUMBER APPROACH

The steps included in the previous section are summarized in the following (Figure 5.4) for quick inspection of the sequence in the solution process. In general, these steps are applicable to analysis of many planar mechanisms by the complex number method.

5.3   METHOD OF UNIT VECTORS

In motion studies of rigid bodies in 3D space, this method is more commonly known as the method of motion relative to a rotating frame of reference (*RFR*). Sometimes it is referred as the method of unit vectors. In principle, there is no apparent advantage of expressing the motion in terms of quantities reference to the fixed or inertial or Newtonian frame of reference (*FFR*), or reference to the *RFR*. For example, in motion analysis of mechanisms that contain several or many rigid bodies one may argue that the *FFR* approach may turn out to be more explicit and algebraically more efficient. However, in many situations the unit vectors or *RFR* approach can simplify the analysis. Specifically, for example, in many rotating machinery problems, such as rotors with flexible shafts, the governing differential equations of motion by the *FFR* approach contain time-dependent coefficients [1]. These are called Hill's equations [2] and are relatively difficult to solve. On the other hand, if the unit vectors or *RFR* approach is adopted the resulting differential equations of motion have constant coefficients and therefore they are relatively much easy to solve. This is the main reason why the *RFR* method or method of unit vectors is introduced in the following.

5.3.1   RATE OF CHANGE OF A VECTOR WITH RESPECT TO A FIXED FRAME

In Figure 5.5 the *FFR* and *RFR* are both centered at the origin O.

Let $\vec{\Omega}$ denote the angular velocity of the frame $Oxyz$ or *RFR* at a given instant (with respect to $OXYZ$).

Consider the position vector

$$\vec{P} = P_x \vec{\imath} + P_y \vec{\jmath} + P_z \vec{k}, \tag{5.19}$$

where P_x, P_y, and P_z are, respectively, the magnitudes of $\vec{P}$ resolving along the unit vectors $\vec{\imath}$, $\vec{\jmath}$, and $\vec{k}$ which are attached to the x, y, and z co-ordinates, respectively. In some textbooks, the latter co-ordinates are referred to as the body fixed co-ordinates (BFC). In this book the unit vectors $\vec{\imath}$, $\vec{\jmath}$, and $\vec{k}$ follow the right-hand screw rule, unless it is stated otherwise.

Differentiating Equation (5.19) with respect to time t and considering the unit vectors ($\vec{\imath}$, $\vec{\jmath}$, and $\vec{k}$) as fixed (with respect to $Oxyz$), then the rate of change of P with respect to the *RFR*

$$\left(\frac{d\vec{P}}{dt}\right)_{Oxyz} = \frac{dP_x}{dt}\vec{\imath} + \frac{dP_y}{dt}\vec{\jmath} + \frac{dP_z}{dt}\vec{k}, \tag{5.20}$$

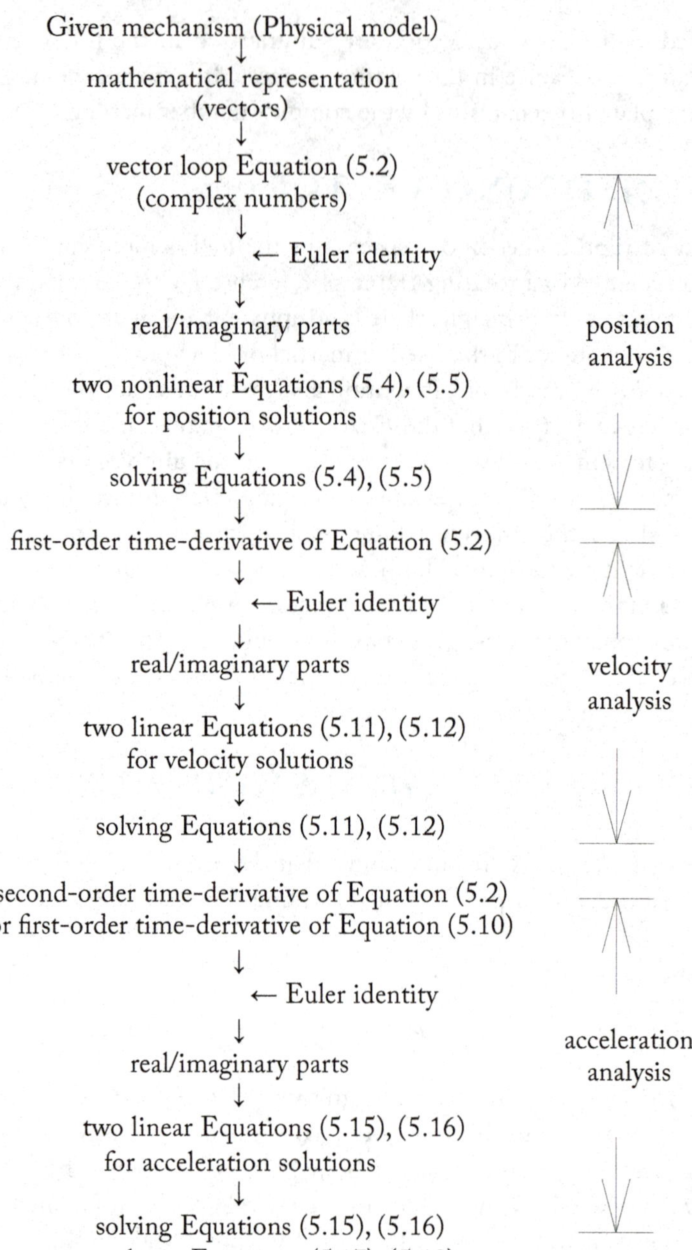

Figure 5.4: Sequence in the solution process.

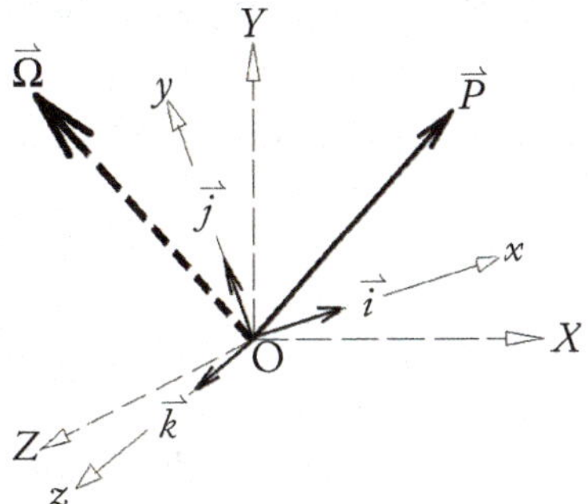

Figure 5.5: A vector in FFR and RFR with same origin.

where the subscript $Oxyz$ on the lhs of Equation (5.20) denotes the *RFR*.

To obtain the rate of change of P with respect to the *FFR*, $\left(\frac{d\vec{P}}{dt}\right)_{OXYZ}$ one must consider the unit vectors ($\vec{\imath}$, $\vec{\jmath}$, and $\vec{k}$) as variables when operating Equation (5.19) because the *RFR* is rotating with respect to the *FFR*. In other words, the movement of *RFR* is time-dependent and thus,

$$\left(\frac{d\vec{P}}{dt}\right)_{OXYZ} = \frac{dP_x}{dt}\vec{\imath} + \frac{dP_y}{dt}\vec{\jmath} + \frac{dP_z}{dt}\vec{k} + P_x\frac{d\vec{\imath}}{dt} + P_y\frac{d\vec{\jmath}}{dt} + P_z\frac{d\vec{k}}{dt}. \qquad (5.21)$$

Applying Equation (5.20) to (5.21), the latter becomes

$$\left(\frac{d\vec{P}}{dt}\right)_{OXYZ} = \left(\frac{d\vec{P}}{dt}\right)_{Oxyz} + P_x\frac{d\vec{\imath}}{dt} + P_y\frac{d\vec{\jmath}}{dt} + P_z\frac{d\vec{k}}{dt}$$

or in more concise form

$$\left(\frac{d\vec{P}}{dt}\right)_{OXYZ} = \left(\frac{d\vec{P}}{dt}\right)_{Oxyz} + \vec{\Omega} \times \vec{P} \qquad (5.22)$$

where the second term on the rhs of Equation (5.22),

$$\vec{\Omega} \times \vec{P} = P_x\frac{d\vec{\imath}}{dt} + P_y\frac{d\vec{\jmath}}{dt} + P_z\frac{d\vec{k}}{dt}$$

because, for example, $\frac{d\vec{\imath}}{dt} = \vec{\Omega} \times \vec{\imath}$, which is the derivative of a unit position vector with respect to time, is the velocity. In other words,

$$P_x\frac{d\vec{\imath}}{dt} + P_y\frac{d\vec{\jmath}}{dt} + P_z\frac{d\vec{k}}{dt} = P_x\vec{\Omega} \times \vec{\imath} + P_y\vec{\Omega} \times \vec{\jmath} + P_z\vec{\Omega} \times \vec{k}.$$

By factoring the common term $\vec{\Omega}\times$ from the rhs of the above equation, it becomes

$$P_x\frac{d\vec{\imath}}{dt} + P_y\frac{d\vec{\jmath}}{dt} + P_z\frac{d\vec{k}}{dt} = \vec{\Omega} \times \left(P_x\vec{\imath} + P_y\vec{\jmath} + P_z\vec{k}\right).$$

5.3.2 MOTION INVOLVED WITH CORIOLIS ACCELERATION

Let P be a particle moving in space as indicated in Figure 5.6. The absolute velocity $\vec{v}_p$ of the particle is defined as the rate of change of $\vec{r}$ with respect to the *FFR*. Therefore, by making use of Equation (5.22), one has

$$\vec{v}_P = \left(\frac{d\vec{r}}{dt}\right)_{OXYZ} = \left(\frac{d\vec{r}}{dt}\right)_{Oxyz} + \vec{\Omega} \times \vec{r}, \tag{5.23}$$

where $\left(\frac{d\vec{r}}{dt}\right)_{Oxyz}$ is the velocity of the particle P relative to the *RFR*.

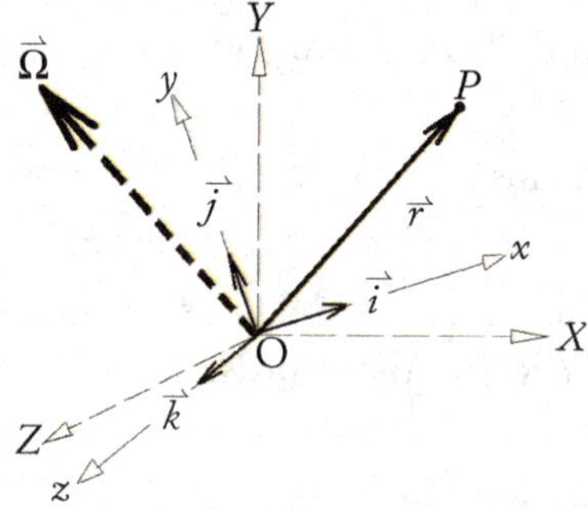

Figure 5.6: A particle in FFR and RFR with same origin.

If one considers the *RFR* as a rigid body, the term $\vec{\Omega} \times \vec{r}$ represents the velocity of a point P' of the *RFR* which coincides with P at the instant considered. Thus,

$$\vec{v}_P = \vec{v}_{P'} + \vec{v}_{P/F} \tag{5.24}$$

where

$$\vec{v}_P = \text{absolute instantaneous velocity of particle } P,$$

$$\vec{v}_{P'} = \text{instantaneous velocity of point } P' \text{of } RFR \text{ coinciding with } P, \text{ and}$$

$$\vec{v}_{P/F} = \text{instantaneous velocity of point } P \text{ relative to the } RFR.$$

In what follows, the adjective "instantaneous" will be disregarded for conciseness. The absolute acceleration $\vec{a}_P$ of the particle P is defined as the change of $\vec{v}_P$ with respect to the *FFR*. Thus, operating on Equation (5.23), one has

$$\vec{a}_P = \left(\frac{d\vec{v}_P}{dt}\right)_{OXYZ}$$

$$= \left[\frac{d}{dt}\left(\frac{d\vec{r}}{dt}\right)_{Oxyz}\right]_{OXYZ} + \frac{d\vec{\Omega}}{dt} \times \vec{r} + \vec{\Omega} \times \left(\frac{d\vec{r}}{dt}\right)_{OXYZ}. \tag{5.25a}$$

Applying Equation (5.22) to the first term on the rhs of the above equation, one has

$$\left[\frac{d}{dt}\left(\frac{d\vec{r}}{dt}\right)_{Oxyz}\right]_{OXYZ} = \left(\frac{d^2\vec{r}}{dt^2}\right)_{Oxyz} + \vec{\Omega}\times\left(\frac{d\vec{r}}{dt}\right)_{Oxyz}$$

and making use of Equation (5.23) for part of the third term on the rhs of Equation (5.25a), that is, $\left(\frac{d\vec{r}}{dt}\right)_{OXYZ} = \left(\frac{d\vec{r}}{dt}\right)_{Oxyz} + \vec{\Omega}\times\vec{r}$, so that the acceleration at point P becomes

$$\vec{a}_P = \left(\frac{d\vec{v}_P}{dt}\right)_{OXYZ}$$

$$= \left(\frac{d^2\vec{r}}{dt^2}\right)_{Oxyz} + \vec{\Omega}\times\left(\frac{d\vec{r}}{dt}\right)_{Oxyz} + \frac{d\vec{\Omega}}{dt}\times\vec{r} + \vec{\Omega}\times\left[\left(\frac{d\vec{r}}{dt}\right)_{Oxyz} + \vec{\Omega}\times\vec{r}\right].$$

Collecting common terms and re-arranging one obtains

$$\vec{a}_P = \left(\frac{d^2\vec{r}}{dt^2}\right)_{Oxyz} + \vec{\alpha}\times\vec{r} + \vec{\Omega}\times(\vec{\Omega}\times\vec{r}) + 2\vec{\Omega}\times\left(\frac{d\vec{r}}{dt}\right)_{Oxyz} \tag{5.25b}$$

in which the angular acceleration $\vec{\alpha} = \frac{d\vec{\Omega}}{dt}$ has been used.

Remarks

- The sum of the second and third terms on the rhs of Equation (5.25b) represents the acceleration $\vec{a}_{p'}$ of the point P' of the rotating frame which coincides with P at the instant considered.

- The first term on the rhs of Equation (5.25b) defines the acceleration of P relative to the rotating frame and is represented by $\vec{a}_{P/F}$.

- The last term on the rhs of Equation (5.25b) is called the complementary acceleration or Coriolis acceleration (after the French mathematician De Coriolis, 1792–1843), denoted by $\vec{a}_c$.

Thus, Equation (5.25b) can be written as

$$\vec{a}_P = \vec{a}_{P/F} + \vec{a}_{P'} + \vec{a}_c. \tag{5.26}$$

- Equations (5.24) and (5.26) may be used to analyze motions of mechanisms which contain parts sliding on each other.

- The concept of Coriolis acceleration is very useful in the study of long range projectiles and other bodies whose motions are appreciably affected by the rotation of the earth.

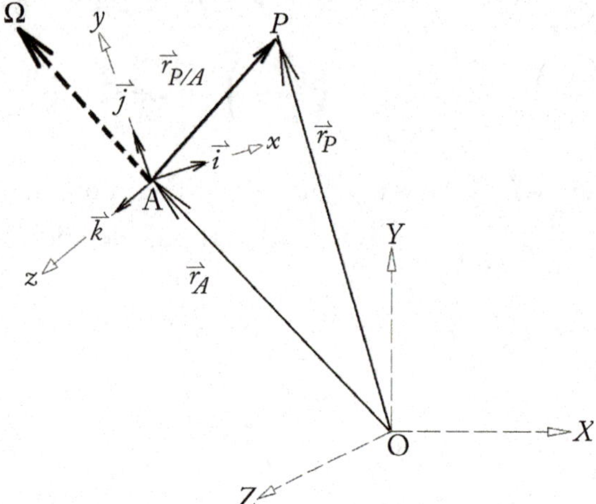

Figure 5.7: A vector in 3D space with FFR and RFR.

5.3.3   GENERAL MOTION INVOLVED MOVING FRAME OF REFERENCE

With reference to Figure 5.7, the position of P is defined at any instant by the vector $\vec{r}_P$ in the *FFR*. Thus, one can write

$$\vec{r}_P = \vec{r}_A + \vec{r}_{P/A}.$$

The velocity at point P becomes

$$\vec{v}_P = \left(\frac{d\vec{r}_P}{dt}\right)_{OXYZ} = \vec{v}_A + \vec{v}_{P/A}.$$

The velocity $\vec{v}_{P/A}$ may be obtained from Equation (5.23) with $\vec{r}$ in the latter equation replaced by $\vec{r}_{P/A}$. Thus,

$$\vec{v}_P = \left(\frac{d\vec{r}_P}{dt}\right)_{OXYZ} = \vec{v}_A + \left(\frac{d\vec{r}_{P/A}}{dt}\right)_{Oxyz} + \vec{\Omega} \times \vec{r}_{P/A}, \tag{5.27}$$

where $\vec{\Omega}$ is the angular velocity of the *RFR* at the instant of interest.

Likewise, the acceleration at point P can be similarly derived or by using Equation (5.25b) to give

$$\vec{a}_P = \left(\frac{d\vec{v}_P}{dt}\right)_{OXYZ} = \vec{a}_A + \left(\frac{d^2\vec{r}_{P/A}}{dt^2}\right)_{Oxyz} + \vec{\alpha} \times \vec{r}_{P/A}$$

$$+ \vec{\Omega} \times (\vec{\Omega} \times \vec{r}_{P/A}) + 2\vec{\Omega} \times \left(\frac{d\vec{r}_{P/A}}{dt}\right)_{Oxyz}. \tag{5.28}$$

Equation (5.28) has five terms on the rhs and is similar to Equation (5.25b) except that in Equation (5.28) there is the additional term $\vec{a}_A$ which is the acceleration of the origin of the *RFR*.

Before leaving this section it should be pointed out that frequently one is wondering how to solve problems by using the above equation, Equation (5.28). In fact, it is not difficult to solve such problems as long as one focuses on the identification of every term in the latter equation and remembering that, in general, every acceleration term has two components, namely, the normal and tangential components. The steps in solving such a problem will be illustrated in Examples 5.2 and 5.3 of the following section.

5.4   QUESTIONS AND SOLUTIONS

In this section three questions and their solutions are presented. The first question is concerned with the application of complex number method. The second and third questions deal with the unit vectors or *RFR* method.

Example 5.1

The mechanism shown in Figure 5.8a has known magnitude of position, angular position, velocity, and acceleration of link 2. They are, respectively, r_2, θ_2, ω_2, and α_2. In addition, the magnitudes of link 3 and offset are given as r_3 and r_d.

By applying the complex number method, find all the unknown expressions for θ_3, $\vec{\omega}_3$, $\vec{\alpha}_3$, $\vec{r}_a$, $\frac{d\vec{r}_a}{dt}$, and $\frac{d^2\vec{r}_a}{dt^2}$.

Solution:

The solution of this question consists of three stages. They are the position solution, velocity solution, and acceleration solution.

(a) **Position solution**

For the above mechanism, draw the vector loop diagram as shown in Figure 5.8b.

With reference to Figures 5.8a and 5.8b, the vector loop equation from the vector loop diagram is given by

$$\vec{r}_2 + \vec{r}_3 = \vec{r}_a + \vec{r}_d. \tag{5.29a}$$

Complex number representation of Equation 5.29a gives

$$r_2 e^{i\theta_2} + r_3 e^{i\theta_3} = r_a e^{i0} + r_d e^{i\frac{\pi}{2}}. \tag{5.29b}$$

Application of Euler identity, also noting $\theta_1 = 0$, Equation (5.29b) becomes

$$r_2\left(\cos\theta_2 + i\sin\theta_2\right) + r_3\left(\cos\theta_3 + i\sin\theta_3\right)$$
$$= r_a\left(\cos 0 + i\sin 0\right) + r_d\left(\cos\pi/2 + i\sin\pi/2\right)$$

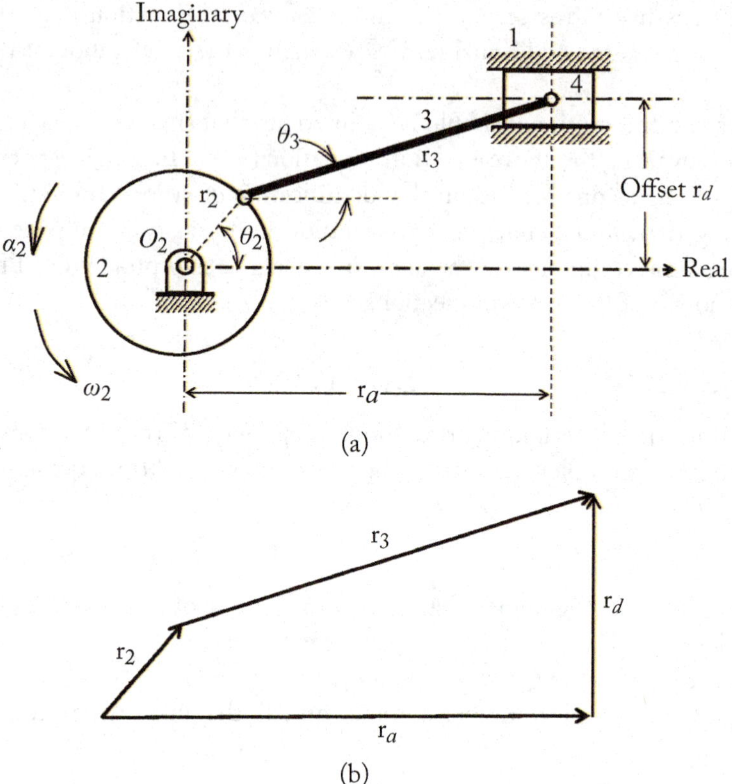

Figure 5.8: (a) Slider crack mechanism; and (b) vector loop diagram of (a).

$$r_2 (\cos \theta_2 + i \sin \theta_2) + r_3 (\cos \theta_3 + i \sin \theta_3) = r_a + i r_d.$$

Separation of real and imaginary parts,

$$r_2 \cos \theta_2 + r_3 \cos \theta_3 = r_a \tag{5.30}$$

$$r_2 \sin \theta_2 + r_3 \sin \theta_3 = r_d. \tag{5.31}$$

The unknowns are r_a and θ_3.

From Equation 5.31, the unknown angle can be found as

$$\theta_3 = \sin^{-1} \left(\frac{r_d - r_2 \sin \theta_2}{r_3} \right). \tag{5.32}$$

Substituting this result into Equation 5.30, one obtains r_a.

(b) **Velocity solution**

Differentiation of Equation (5.29b) with respect to time t, gives

$$\frac{dr_a}{dt} + i\frac{dr_d}{dt} = r_2\left(\frac{d\theta_2}{dt}\right)\left(ie^{i\theta_2}\right) + \left(\frac{dr_2}{dt}\right)e^{i\theta_2}$$
$$+ r_3\left(\frac{d\theta_3}{dt}\right)\left(ie^{i\theta_3}\right) + \left(\frac{dr_3}{dt}\right)e^{i\theta_3}.$$

Note that r_d, r_2, and r_3 are constant and therefore their derivatives with respect to time t are zero so that the above equation reduces to

$$\frac{dr_a}{dt} = r_2\left(\frac{d\theta_2}{dt}\right)\left(ie^{i\theta_2}\right) + r_3\left(\frac{d\theta_3}{dt}\right)\left(ie^{i\theta_3}\right).$$

Application of Euler identity, it becomes

$$\frac{dr_a}{dt} = r_2\omega_2\left[i\left(\cos\theta_2 + i\sin\theta_2\right)\right]$$
$$+ r_3\omega_3\left[i\left(\cos\theta_3 + i\sin\theta_3\right)\right]$$
$$= ir_2\omega_2\cos\theta_2 - r_2\omega_2\sin\theta_2$$
$$+ ir_3\omega_3\cos\theta_3 - r_3\omega_3\sin\theta_3.$$

Separation of real and imaginary parts,

$$\frac{dr_a}{dt} = -r_2\omega_2\sin\theta_2 - r_3\omega_3\sin\theta_3, \tag{5.33}$$

$$0 = r_2\omega_2\cos\theta_2 + r_3\omega_3\cos\theta_3. \tag{5.34}$$

From Equation (5.34), one has

$$\omega_3 = \frac{-r_2\omega_2\cos\theta_2}{r_3\cos\theta_3}. \tag{5.35}$$

Substituting Equation (5.35) into (5.33), one obtains

$$\frac{dr_a}{dt} = -r_2\omega_2\sin\theta_2 - r_3\left(\frac{-r_2\omega_2\cos\theta_2}{r_3\cos\theta_3}\right)\sin\theta_3,$$

$$\frac{dr_a}{dt} = r_2\omega_2\left(\cos\theta_2\tan\theta_3 - \sin\theta_2\right). \tag{5.36}$$

Equations (5.35) and (5.36) are the velocity solutions.

(c) Acceleration solution

Differentiation of Equation (5.29b) twice with respect to time t, gives

$$\frac{d^2 r_a}{dt^2} = r_2 \left[\left(\frac{d^2 \theta_2}{dt^2} \right) \left(i e^{i\theta_2} \right) + \omega_2^2 i^2 e^{i\theta_2} \right]$$

$$+ r_3 \left[\left(\frac{d^2 \theta_3}{dt^2} \right) \left(i e^{i\theta_3} \right) + \omega_3^2 i^2 e^{i\theta_3} \right]$$

$$\frac{d^2 r_a}{dt^2} = r_2 \left[\left(\frac{d^2 \theta_2}{dt^2} \right) \left(i e^{i\theta_2} \right) - \omega_2^2 e^{i\theta_2} \right]$$

$$+ r_3 \left[\left(\frac{d^2 \theta_3}{dt^2} \right) \left(i e^{i\theta_3} \right) - \omega_3^2 e^{i\theta_3} \right]$$

$$= r_2 \left[\left(\frac{d^2 \theta_2}{dt^2} \right) \left(i e^{i\theta_2} \right) - \omega_2^2 e^{i\theta_2} \right]$$

$$+ r_3 \left[\left(\frac{d^2 \theta_3}{dt^2} \right) \left(i e^{i\theta_3} \right) - \omega_3^2 e^{i\theta_3} \right].$$

Writing $\frac{d^2 \theta_2}{dt^2} = \alpha_2$ and $\frac{d^2 \theta_3}{dt^2} = \alpha_3$, and application of Euler's identity, it becomes

$$\frac{d^2 r_a}{dt^2} = i \left[r_2 \alpha_2 \left(\cos \theta_2 + i \sin \theta_2 \right) + r_3 \alpha_3 \left(\cos \theta_3 + i \sin \theta_3 \right) \right]$$

$$- r_2 \omega_2^2 \left(\cos \theta_2 + i \sin \theta_2 \right) - r_3 \omega_3^2 \left(\cos \theta_3 + i \sin \theta_3 \right)$$

$$= i r_2 \alpha_2 \cos \theta_2 - r_2 \alpha_2 \sin \theta_2 + i r_3 \alpha_3 \cos \theta_3 - r_3 \alpha_3 \sin \theta_3$$

$$- r_2 \omega_2^2 \cos \theta_2 - i r_2 \omega_2^2 \sin \theta_2 - r_3 \omega_3^2 \cos \theta_3 - i r_3 \omega_3^2 \sin \theta_3$$

$$= -r_2 \omega_2^2 \cos \theta_2 - r_2 \alpha_2 \sin \theta_2 - r_3 \omega_3^2 \cos \theta_3 - r_3 \alpha_3 \sin \theta_3$$

$$+ i \left(r_2 \alpha_2 \cos \theta_2 - r_2 \omega_2^2 \sin \theta_2 + r_3 \alpha_3 \cos \theta_3 - r_3 \omega_3^2 \sin \theta_3 \right).$$

Separation of real and imaginary parts,

$$\frac{d^2 r_a}{dt^2} = -r_2 \omega_2^2 \cos \theta_2 - r_2 \alpha_2 \sin \theta_2 - r_3 \omega_3^2 \cos \theta_3 - r_3 \alpha_3 \sin \theta_3 \tag{5.37}$$

$$0 = r_2 \alpha_2 \cos \theta_2 - r_2 \omega_2^2 \sin \theta_2 + r_3 \alpha_3 \cos \theta_3 - r_3 \alpha_3^2 \sin \theta_3. \tag{5.38}$$

The unknowns are $\frac{d^2 r_a}{dt^2}$ and α_3. From Equation (5.38),

$$\alpha_3 = \frac{r_2 \omega_2^2 \sin \theta_2 + r_3 \omega_3^2 \sin \theta_3 - r_2 \alpha_2 \cos \theta_2}{r_3 \cos \theta_3}$$

$$= \omega_3^2 \tan \theta_3 + \left(\frac{r_2}{r_3} \right) \left(\frac{\omega_2^2 \sin \theta_2 - \alpha_2 \cos \theta_2}{\cos \theta_3} \right). \tag{5.39}$$

Substituting Equation (5.39) into (5.37), one arrives at

$$\frac{d^2 r_a}{dt^2} = -r_2 \omega_2^2 \cos \theta_2 - r_2 \alpha_2 \sin \theta_2 - r_3 \omega_3^2 \cos \theta_3$$

$$- r_3 \sin \theta_3 \left[\omega_3^2 \tan \theta_3 + \left(\frac{r_2}{r_3} \right) \left(\frac{\omega_2^2 \sin \theta_2 - \alpha_2 \cos \theta_2}{\cos \theta_3} \right) \right]. \tag{5.40}$$

Equations (5.39) and (5.40) are the acceleration solution.

Example 5.2
The cam mechanism shown in Figure 5.9a has a constant angular velocity $\omega_2 = 2.0$ rad/s clockwise. The diameter of the roller (link 3) of the follower is 12.7 mm. The dimensions in this figure are in mm. By applying the method of unit vectors, determine the velocity and acceleration at point A of the follower (more precisely, follower shaft, link 4) for the phase shown. Given that the angle O_2AB is 64.47° (note that O_2B is parallel to the y-axis of the RFR). This problem with a different angular velocity ω_2 and orientation of the follower shaft was provided by Mabie and Ocvirk [3].

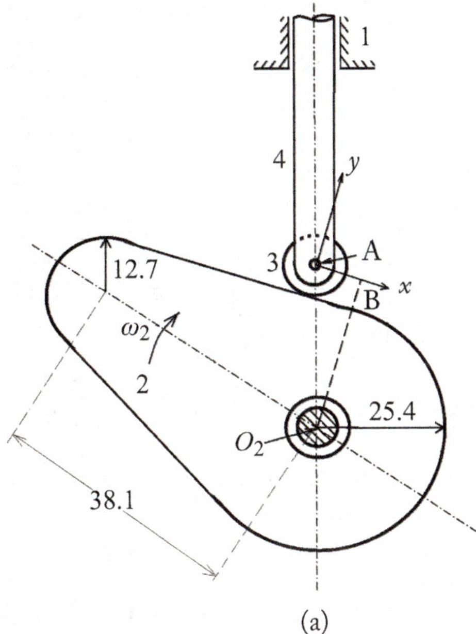

Figure 5.9: (a) Cam mechanism (dimensions in mm).

Solution:

Let O_2 be the origem of the *FFR* and A be the origin of the *RFR* with x-axis parallel to the flat face of the cam as shown in the sketch in Figure 5.9b. Imagine A is attached to link 2 then A_2 denotes the point A associated with link 2. The point A_4 of interest is attached to the shaft of the follower (that is, A_4 is the point A associated with link 4). The points A_2 and A_4 are the so-called co-incident points.

(a) *Position Analysis*

With reference to Figure 5.9b,

$CD = 12.7 + 12.7/2$ mm $= 19.05$ mm.

$O_2B = 25.4 + 12.7/2$ mm $= 31.75$ mm.

$O_2C = 38.1$ mm.

$O_2E = O_2B - CD = 12.7$ mm.

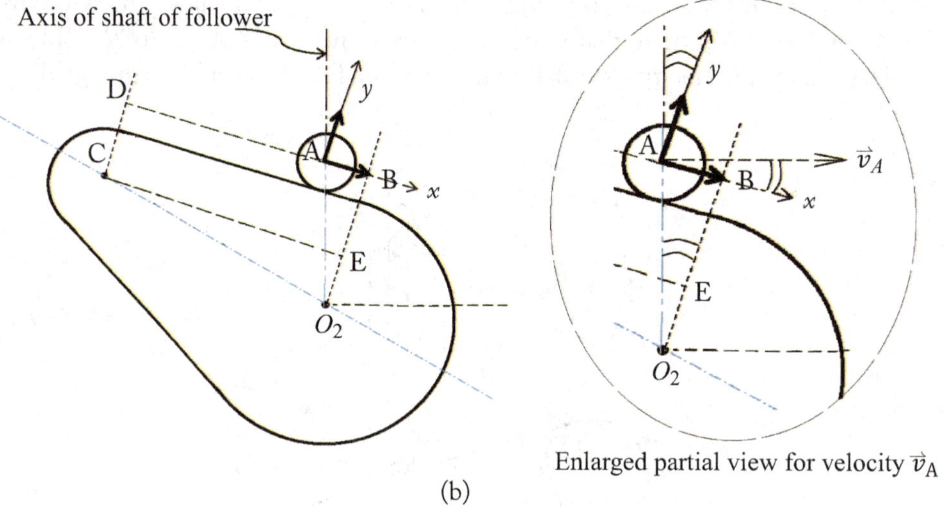

Enlarged partial view for velocity $\vec{v}_A$

(b)

Figure 5.9: (b) Sketeches of cam mechanism for position and velocity analyses.

Since $\angle O_2AB = 64.47°$, given, and therefore $\angle AO_2B = 90° - 64.47° = 25.53°$. Thus, $O_2A = O_2B/\cos AO_2B = 35.186$ mm.

(b) *Velocity Analysis*

From Equation (5.27), and note that the point of interest P is A_4 and A is A_2 in the present problem, the velocity of interest is

$$\vec{v}_P = \vec{v}_{A_4} = \left(\frac{d\vec{r}_P}{dt}\right)_{OXYZ} = \vec{v}_{A_2} + \left(\frac{d\vec{r}_{A_4/A_2}}{dt}\right)_{Oxyz} + \vec{\Omega} \times \vec{r}_{A_4/A_2}. \qquad (5.41)$$

The terms in Equation (5.41) are now identified.

Since the follower can only move along the axis of its shaft and with reference to Figures 5.9a and 5.9b,

$$\vec{v}_P = v_P \left(-\sin 25.53°\,\vec{\imath} + \cos 25.53°\,\vec{\jmath}\right)$$
$$= v_P \left(-0.4310\,\vec{\imath} + 0.9024\,\vec{\jmath}\right).$$

The magnitude of the velocity v_P is unknown while the direction is along the axis of its shaft as indicated in the foregoing.

The velocity at A is perpendicular to O_2A (recall, the velocity at a point is always perpendicular to the position vector) and its magnitude is known since the length O_2A and ω_2 are determined and given, respectively. Since A_2 is associated with link 2, thus line O_2A can be considered as rotating with the angular velocity of link 2, $\vec{\omega}_2$ which is $\vec{\Omega}$ in Equation (5.41). Hence,

$$\vec{v}_A = \vec{v}_{A_2} = \omega_2\left(O_2A\right)\left(\cos 25.53°\,\vec{\imath} + \sin 25.53°\,\vec{\jmath}\right)$$
$$= 2(35.186)\left(\cos 25.53°\,\vec{\imath} + \sin 25.53°\,\vec{\jmath}\right) \qquad \text{mm/s}$$
$$= 63.50\,\vec{\imath} + 30.33\,\vec{\jmath} \qquad \text{mm/s.}$$

The relative velocity of P with respect to the origin of the *RFR* referring to the *RFR* is

$$\left(\frac{d\vec{r}_{P/A}}{dt}\right)_{Oxyz} = v\,\vec{\imath}.$$

That is, in general, the relative velocity at P with respect to A is not zero as P and A are associated with different links. One may wonder why the above equation holds true. The implicit assumption in this problem is that the roller of the follower does not separate from the surface of the cam and it does not penetrate into the cam. In other words, the only relative motion possible is along the surface of the cam.

However,

$$\vec{\Omega} \times \vec{r}_{P/A} = 0, \qquad \text{since} \qquad \vec{r}_{P/A} = 0.$$

Substituting all the terms into Equation (5.41), one has

$$v_P\left(-0.4310\,\vec{\imath} + 0.9024\,\vec{\jmath}\right) = 63.50\,\vec{\imath} + 30.33\,\vec{\jmath} + v\,\vec{\imath}. \qquad (5.42)$$

Equating coefficients of terms associated with $\vec{\jmath}$ results

$$v_P(0.9024) = 30.33,$$

giving $v_P = 33.61$ mm/s .

Substituting into Equation (5.42) and equating coefficients of terms associated with $\vec{\imath}$, one finds $v = -77.986$ mm/s. Therefore, the required velocity is

$$\vec{v}_P = -14.486\,\vec{\imath} + 30.33\,\vec{\jmath} \text{ mm/s.}$$

(c) *Acceleration Analysis*

The acceleration from Equation (5.28) is

$$\vec{a}_P = \vec{a}_A + \left(\frac{d^2 \vec{r}_{P/A}}{dt^2}\right)_{Axyz}$$

$$+ \vec{\alpha} \times \vec{r}_{P/A} + \vec{\Omega} \times (\vec{\Omega} \times \vec{r}_{P/A}) + 2\vec{\Omega} \times \left(\frac{d\vec{r}_{P/A}}{dt}\right)_{Axyz}. \tag{5.43}$$

Since $\vec{r}_{P/A} = 0$, therefore Equation (5.43) reduces to

$$\vec{a}_P = \vec{a}_A + \left(\frac{d^2 \vec{r}_{P/A}}{dt^2}\right)_{Axyz} + 2\vec{\Omega} \times \left(\frac{d\vec{r}_{P/A}}{dt}\right)_{Axyz}. \tag{5.44}$$

Consider first the lhs term,

$$\vec{a}_P = (\vec{a}_P)_n + (\vec{a}_P)_t$$

in which the normal component $(\vec{a}_P)_n = 0$ because the follower is supposed to move along the axis of the shaft only. Otherwise, the follower would not move because the non-zero normal component (perpendicular to the axis of the shaft of the follower) of the acceleration at P will result in a couple that will jam the follower. Thus, the tangential component (along the axis of the shaft of the follower; also parallel to the velocity at P) is

$$(\vec{a}_P)_t = (a_P)_t \left(-\sin 25.53° \vec{\imath} + \cos 25.53° \vec{\jmath}\right)$$
$$= (a_P)_t \left(-0.4310 \vec{\imath} + 0.9024 \vec{\jmath}\right).$$

The first term on the rhs of Equation (5.44) is

$$\vec{a}_A = \omega_2^2 \left(O_2 A\right) \left(\sin 25.53° \vec{\imath} - \cos 25.53° \vec{\jmath}\right)$$
$$= 60.66 \vec{\imath} - 127.0 \vec{\jmath} \quad \text{mm/s}^2.$$

This is the normal component and it is along the axis of the shaft of the follower. The direction is from A toward O_2 (recall, the normal component of the acceleration at a point is always directing toward the center of curvature in a circular motion). The tangential component is zero since the given angular velocity $\vec{\omega}_2$ is constant. Physically, if this tangential component is not zero it would means a force along this direction will result a moment jamming the shaft of the follower. In order to avoid jamming the shaft of the follower the tangential component of this acceleration should be zero and therefore the angular velocity $\vec{\omega}_2$ has to be constant.

The second term on the rhs of Equation (5.44) is

$$\left(\frac{d^2 \vec{r}_{P/A}}{dt^2}\right)_{Axyz} = \left(\frac{d^2 \vec{r}_{P/A}}{dt^2}\right)_{Axyz}^t + \left(\frac{d^2 \vec{r}_{P/A}}{dt^2}\right)_{Axyz}^n = \left(\frac{d^2 \vec{r}_{P/A}}{dt^2}\right)_{Axyz}^t = a\vec{\imath},$$

where the superscript t and n denote, respectively, the tangential and normal components of the acceleration. The magnitude of the acceleration a is unknown. Since only the relative acceleration parallel (that is, tangential) to the direction of the flat surface of the cam is possible (that is, the normal component is zero; otherwise, the roller of the follower will leave or lose contact with the surface of the cam or penetrate into the surface which is impossible as all the links are assumed to be rigid).

The third term on the rhs of Equation (5.44) is

$$2\vec{\Omega} \times \left(\frac{d\vec{r}_{P/A}}{dt}\right)_{Axyz} = 2\left(-\omega_2\vec{k}\right) \times v\vec{\imath} = 2\left(-\omega_2\vec{k}\right) \times (-77.986)\vec{\imath}$$

$$= 311.944\,\vec{\jmath} \qquad \text{mm/s}^2.$$

Substituting all of the above terms into Equation (5.44), one obtains

$$\left(a_p\right)_t \left(-0.4310\,\vec{\imath} + 0.9024\,\vec{\jmath}\right) = 60.66\,\vec{\imath} - 127.0\,\vec{\jmath} + a\,\vec{\imath} + 311.944\,\vec{\jmath}.$$

Equating coefficients associated with unit vector $\vec{\jmath}$, one has

$$\left(a_p\right)_t (0.9024) = -127.0 + 311.944$$

which gives

$$\left(a_p\right)_t = a_P = 204.947 \qquad \text{mm/s}^2.$$

Thus, the required acceleration of the follower at A_4 is

$$\vec{a}_P = \vec{a}_{A_4} = -88.33\,\vec{\imath} + 184.94\,\vec{\jmath} \qquad \text{mm/s}^2.$$

Remarks:

In this example and that which follows, application of the unit vector method based on Equations (5.27) and (5.28) has been made. In applying these equations to the examples, the terms on the lhs are referenced to the *RFR*. However, in the derivation of Equations (5.27) and (5.28) the terms on the lhs are referenced to the *FFR*. This, naturally, leads to the question of the validity of the solution. To show that whether the terms on the lhs of the equations referenced to the *RFR* or *FFR* are equal the following simple illustration is in order.

Consider the desired velocity $\vec{v}_P$ of the example in terms of the unit vectors $\vec{I}, \vec{J}$, and $\vec{K}$ of the *FFR*. Note that the origin of the *FFR* is at O_2, X-axis is parallel to $\vec{v}_A$, Y-axis is along the axis of the shaft of follower, and Z-axis is perpendicular to the XY-plane. Thus,

$$\vec{v}_P = v_P\,\vec{J}.$$

With reference to Figure 5.9b, the unit vector $\vec{J}$ of the *FFR* is related to the unit vectors of the *RFR* by

$$\vec{J} = -\sin 25.53°\,\vec{i} + \cos 25.53°\,\vec{j}.$$

Therefore, the desired velocity

$$\vec{v}_P = v_P\left(-\sin 25.53°\,\vec{i} + \cos 25.53°\,\vec{j}\right).$$

This equation for $\vec{v}_P$ is identical to that referenced to the *RFR*.

Example 5.3

The four-bar linkage shown in Figure 5.10 has a constant angular velocity $\omega_2 = 1.0$ rad/s and direction as shown. By applying the unit vectors method, determine the velocity and acceleration at point B. Note that the *RFR* has its origin at point A and unit vector $\vec{i}$ attached to link 3 as shown. Thus, the unit vector $\vec{k}$ points into the plane of the figure.

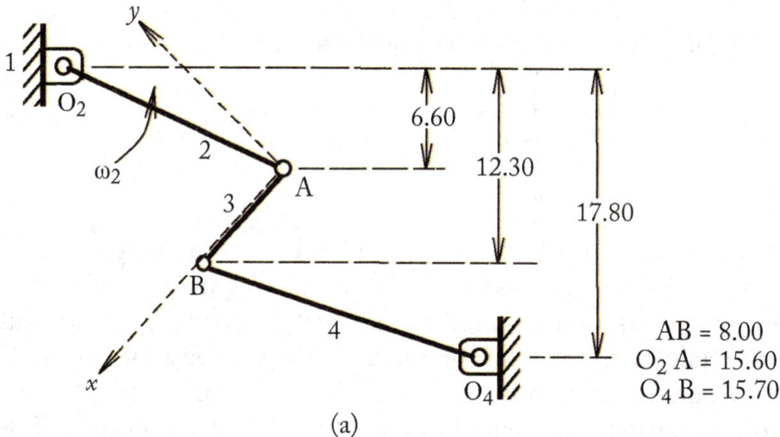

Figure 5.10: Four-bar linkage (dimensions in cm).

Solution:

Let O_2 be the origin of the *FFR*. The velocities at points A and B are indicated in Figure 5.11 in which the angle $\beta, \gamma, \delta, \phi$, and θ are to be determined in the following position analysis.

(a) *Position analysis*

With reference to Figure 5.11 and note that all dimensions are in cm,

$$\sin \gamma = \frac{12.3 - 6.6}{AB} = \frac{12.3 - 6.6}{8.0} = 0.7125 \qquad \text{so that} \quad \gamma = 45.4387°,$$

$$\sin \phi = \frac{17.8 - 12.3}{O_4B} = \frac{17.8 - 12.3}{15.7} = 0.3503 \qquad \text{so that} \quad \phi = 20.50°,$$

$$\sin \theta = \frac{6.6}{O_2A} = \frac{6.6}{15.6} = 0.4231 \qquad \text{so that} \quad \theta = 25.03°,$$

and therefore,

$$\delta = \gamma + \theta = 70.4687°, \qquad \beta = 90° - \delta = 19.53°,$$

and

$$\zeta = 90° - (\gamma + \phi) = 90° - 65.9387° = 24.06°.$$

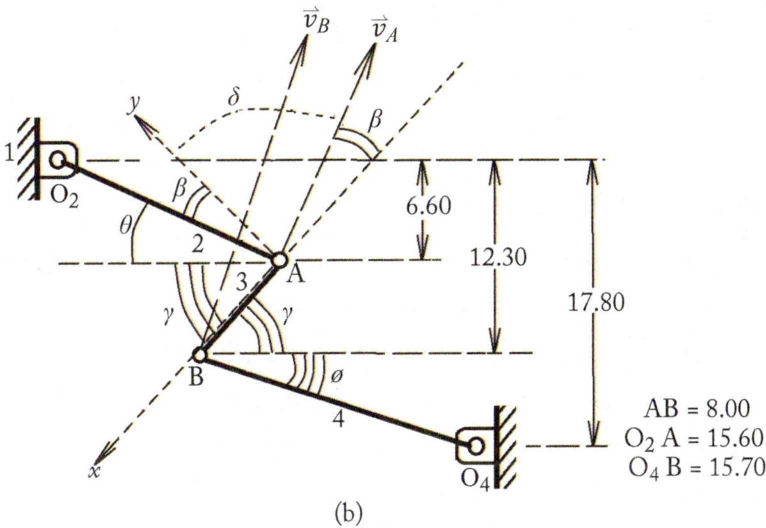

Figure 5.11: Four-bar linkage with angles and velocities (dimensions in cm).

(b) *Velocity analysis*

From Equation (5.27), the velocity of interest is

$$\vec{v}_B = \left(\frac{d\vec{r}_B}{dt}\right)_{OXYZ} = \vec{v}_A + \left(\frac{d\vec{r}_{B/A}}{dt}\right)_{Oxyz} + \vec{\Omega} \times \vec{r}_{B/A}.$$

But points A and B are on the same link, link 3, and therefore the second term on the rhs of the above equation is zero. Thus,

$$\vec{v}_B = \left(\frac{d\vec{r}_B}{dt}\right)_{OXYZ} = \vec{v}_A + \vec{\Omega} \times \vec{r}_{B/A}. \tag{5.45}$$

Now, the terms in Equation (5.45) are identified in the following.

With reference to Figure 5.11, the velocity at point B,

$$\vec{v}_B = v_B\left(-\cos\zeta\,\vec{\imath} + \sin\zeta\,\vec{\jmath}\right).$$

The velocity at A is perpendicular to O_2A (recall, the velocity at a point is always perpendicular to the position vector) and therefore according to Figure 5.11 in which $\vec{\omega}_2$ is given,

$$\vec{v}_A = \vec{v}_{A_2} = \omega_2\left(O_2A\right)\left(-\cos\beta\,\vec{\imath} + \sin\beta\,\vec{\jmath}\right).$$

The second term on the rhs of Equation (5.45) is

$$\vec{\Omega} \times \vec{r}_{B/A} = \omega_3\vec{k} \times r_{B/A}\,\vec{\imath} = \omega_3\left(BA\right)\vec{\jmath}$$

since the *RFR* is attached to link 3 such that $\vec{\Omega} = \omega_3\vec{k}$.

Substituting all of the terms into Equation (5.45), one has

$$v_B\left(-\cos\zeta\,\vec{\imath} + \sin\zeta\,\vec{\jmath}\right) = \omega_2\left(O_2A\right)\left(-\cos\beta\,\vec{\imath} + \sin\beta\,\vec{\jmath}\right) + \omega_3(BA)\vec{\jmath}. \tag{5.46}$$

Equating coefficients of $\vec{\imath}$, one has

$$-v_B\cos\zeta = -\omega_2\left(O_2A\right)\cos\beta$$

which gives

$$v_B = \frac{\omega_2(O_2A)\cos\beta}{\cos\zeta} = \frac{1.0(15.6)\cos 19.53°}{\cos 24.06°}\ \text{cm/s}$$

$$= \frac{1.0(15.6)(0.9425)}{0.9131}\ \text{cm/s} = 16.1023\ \text{cm/s}.$$

Thus, the required velocity,

$$\vec{v}_B = v_B\left(-\cos\zeta\,\vec{\imath} + \sin\zeta\,\vec{\imath}\right) = 16.1023(-0.9131\,\vec{\imath} + 0.4077\,\vec{\jmath})\ \text{cm/s}$$

$$\vec{v}_B = -14.7030\,\vec{\imath} + 6.5649\,\vec{\jmath}\ \text{cm/s}.$$

For Equation (5.46), equating coefficients of $\vec{J}$, it gives

$$v_B \sin \zeta = \omega_2 \left(O_2A\right) \sin \beta + \omega_3 (BA) \qquad \text{so that}$$

$$\begin{aligned}
\omega_3 &= \frac{16.1023 \sin \zeta - 1.0(15.6) \sin \beta}{BA} \quad \text{rad/s}\\
&= \frac{16.1023 \sin 24.06° - 1.0(15.6) \sin 19.53°}{8.0} \quad \text{rad/s}\\
&= \frac{16.1023(0.4077) - 1.0(15.6)(0.3343)}{8.0} \quad \text{rad/s}\\
\omega_3 &= \frac{6.5649 - 5.2151}{8.0} \frac{\text{rad}}{\text{s}} = 0.1687 \quad \text{rad/s}.
\end{aligned}$$

(c) *Acceleration analysis*

The acceleration from Equation (5.28) is

$$\vec{a}_B = \vec{a}_A + \left(\frac{d^2 \vec{r}_{B/A}}{dt^2}\right)_{Axyz}$$

$$+ \vec{\alpha} \times \vec{r}_{B/A} + \vec{\Omega} \times \left(\vec{\Omega} \times \vec{r}_{B/A}\right) + 2\vec{\Omega} \times \left(\frac{d \vec{r}_{B/A}}{dt}\right)_{Axyz}. \tag{5.47}$$

In Equation (5.47), the terms $\left(\frac{d \vec{r}_{B/A}}{dt}\right)_{Axyz} = 0$ and $\left(\frac{d^2 \vec{r}_{B/A}}{dt^2}\right)_{Axyz} = 0$ because points B and A are on the same link, link 3. Therefore, Equation (5.47) reduces to

$$\vec{a}_B = \vec{a}_A + \vec{\alpha} \times \vec{r}_{B/A} + \vec{\Omega} \times \left(\vec{\Omega} \times \vec{r}_{B/A}\right). \tag{5.48}$$

Now, every term in Equation (5.48) is identified in the following.

First, the lhs term

$$\vec{a}_B = (\vec{a}_B)_n + (\vec{a}_B)_t$$

in which the normal component (recall, the normal component is directed from B toward the center of curvature, O_4)

$$\begin{aligned}
(\vec{a}_B)_n &= \frac{v_B^2}{O_4B} \left[-\cos\left(\gamma + \phi\right) \vec{\imath} - \sin\left(\gamma + \phi\right) \vec{J}\right]\\
&= -\frac{16.1023^2}{15.7} \left(\cos 65.9387° \vec{\imath} + \sin 65.9387° \vec{J}\right) \quad \text{cm/s}^2\\
&= -16.5149 \left(0.4077 \vec{\imath} + 0.9131 \vec{J}\right) \quad \text{cm/s}^2\\
&= -\left(6.7331 \vec{\imath} + 15.0798 \vec{J}\right) \quad \text{cm/s}^2
\end{aligned}$$

and the tangential component (which is perpendicular to the normal component or parallel to the velocity $\vec{v}_B$) is

$$(\vec{a}_B)_t = (a_B)_t \left(-\cos \zeta \, \vec{\imath} + \sin \zeta \, \vec{\jmath}\right)$$
$$= (a_B)_t \left(-\cos 24.06° \, \vec{\imath} + \sin 24.06° \, \vec{\jmath}\right)$$
$$(\vec{a}_B)_t = (a_B)_t \left(-0.9131 \, \vec{\imath} + 0.4077 \, \vec{\jmath}\right).$$

Second, the first term on the rhs of Equation (5.48) is considered. It is

$$\vec{a}_A = (\vec{a}_A)_n + (\vec{a}_A)_t$$

in which the tangential component is zero as $\alpha_2 = 0$ because ω_2 is constant. Therefore, $\vec{a}_A = (\vec{a}_A)_n$ and with reference to Figure 5.11, one obtains

$$\vec{a}_A = (\vec{a}_A)_n = \omega_2^2 \, (O_2A) \left(\cos \delta \, \vec{\imath} + \sin \delta \, \vec{\jmath}\right)$$
$$= 15.6 \left(\cos 70.4687° \, \vec{\imath} + \sin 70.4687° \, \vec{\jmath}\right) \qquad \text{cm/s}^2$$
$$= 15.6 \left(0.3343 \, \vec{\imath} + 0.9425 \, \vec{\jmath}\right) \qquad \text{cm/s}^2$$
$$= 5.2151 \, \vec{\imath} + 14.7025 \, \vec{\jmath} \qquad \text{cm/s}^2.$$

Third, the second term on the rhs of Equation (5.48) is considered,

$$\vec{\alpha} \times \vec{r}_{B/A} = \alpha_3 \vec{k} \times r_{B/A} \, \vec{\imath} = \alpha_3 (BA) \, \vec{\jmath},$$

and the third term on the rhs of Equation (5.48) is

$$\vec{\Omega} \times \left(\vec{\Omega} \times \vec{r}_{B/A}\right) = \omega_3 \vec{k} \times \left(\omega_3 \vec{k} \times r_{B/A} \, \vec{\imath}\right) = \omega_3 \vec{k} \times \omega_3 (BA) \, \vec{\jmath}$$
$$= -\omega_3^2 (BA) \, \vec{\imath}.$$

Substituting all of the above terms into Equation (5.48), one obtains

$$- \left(6.7331 \, \vec{\imath} + 15.0798 \, \vec{\jmath}\right) + (a_B)_t \left(0.9131 \, \vec{\imath} + 0.4077 \, \vec{\jmath}\right)$$
$$= 5.2151 \, \vec{\imath} + 14.7025 \, \vec{\jmath} + \alpha_3 (BA) \, \vec{\jmath} - \omega_3^2 (BA) \, \vec{\imath}.$$

Equating coefficients associated with unit vector $\vec{\imath}$, one has

$$-6.7331 - 0.9131 \, (a_B)_t = 5.2151 - 0.2277$$

which gives

$$(a_B)_t = -12.8359 \qquad \text{cm/s}^2.$$

Equating coefficients associated with unit vector $\vec{\jmath}$, it leads to

$$-15.0798 + 0.4077 \, (a_B)_t = 14.7025 + 8.0\alpha_3$$

so that

$$\alpha_3 = -4.3769 \quad \text{rad/s}^2.$$

The required acceleration is

$$\vec{a}_B = (\vec{a}_B)_n + (\vec{a}_B)_t$$
$$= -(6.7331\,\vec{\imath} + 15.0798\,\vec{\jmath}) - 12.8359\,(-0.9131\,\vec{\imath} + 0.4077\,\vec{\jmath}) \quad \text{cm/s}^2$$
$$= 4.9874\,\vec{\imath} - 20.3130\,\vec{\jmath} \quad \text{cm/s}^2.$$

5.5   EXERCISES

5.1.   A mechanism known as Scotch yoke is shown in Figure 5.12. The constant angular velocity of the crank or link 2 is given as ω_2 ccw, indicated in the figure. By applying the complex number method, the vector equation $\vec{R}_{B_4} = \vec{R}_2 + \vec{R}_a$, where $\vec{R}_a$ is the vector measuring from point A to B_4, and assuming the length of link 2, r_2 (that is, the magnitude of $\vec{R}_2$) as well as angle θ_2 are known, find the velocity and acceleration at B_4.

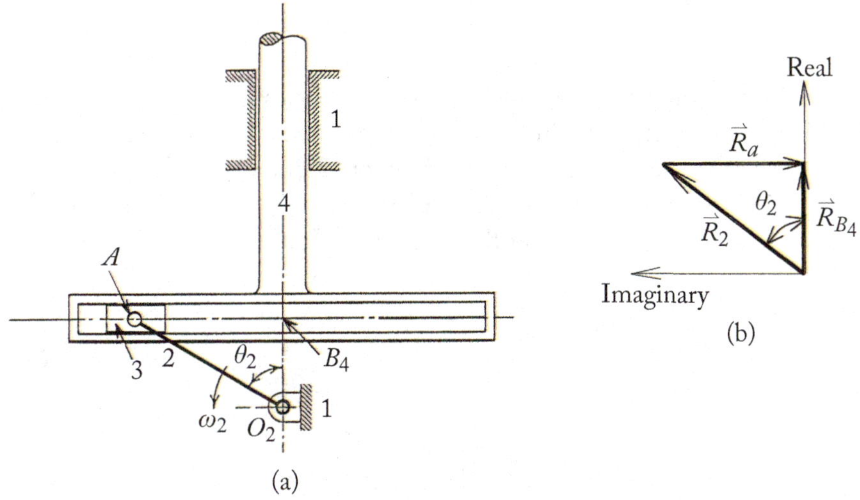

Figure 5.12: Scotch yoke mechanism.

5.2.   The mechanism shown in Figure 5.13 is known as Rapson's slide which has been employed as a marine steering device. Link 2 is called the tiller while link 4, (that is, AC) is the actuating rod. Angle $\theta_2 = 120°$ and the velocity at A associated with link 4 is $\vec{v}_{A_4} = 0.30$ m/s in the direction shown. Note that $\vec{v}_{A_4} = \frac{d\vec{r}_4}{dt}$, and $\vec{v}_{A_2} = \frac{d\vec{r}_2}{dt}$, for example, are used. By applying the complex number method, determine $\frac{d\vec{r}_2}{dt}$, $\vec{\omega}_2$, $\frac{d^2\vec{r}_2}{dt^2}$, and $\vec{\alpha}_2$.

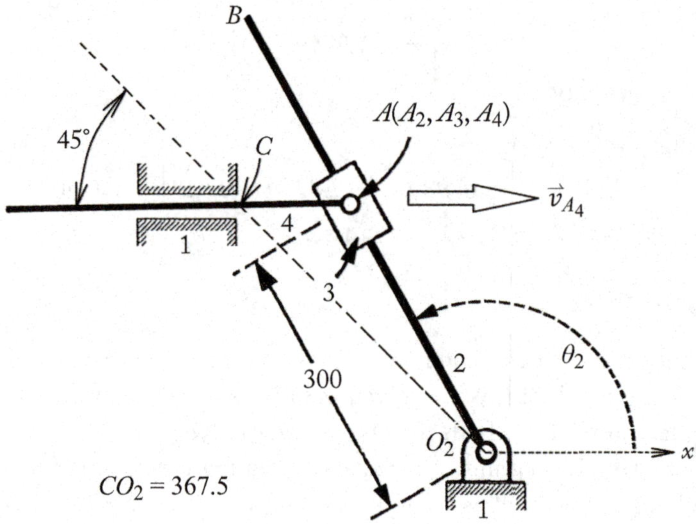

Figure 5.13: Rapson's slide (dimensions in mm).

5.3. For the linkage shown in Figure 5.14 in which the axes of the *RFR* are indicated with its origin at A and attached to link 3, link 2 rotates clockwise at a constant angular velocity, $\omega_2 = 1.0$ rad/s. The angle between the horizontal line O_2O_4 and link 2 is $\theta_2 = 35.51°$. By applying the unit vectors method, determine the angular velocities $\vec{\omega}_3, \vec{\omega}_4$, velocity and acceleration at C. Note that the points O_2, A, and C are collinear.

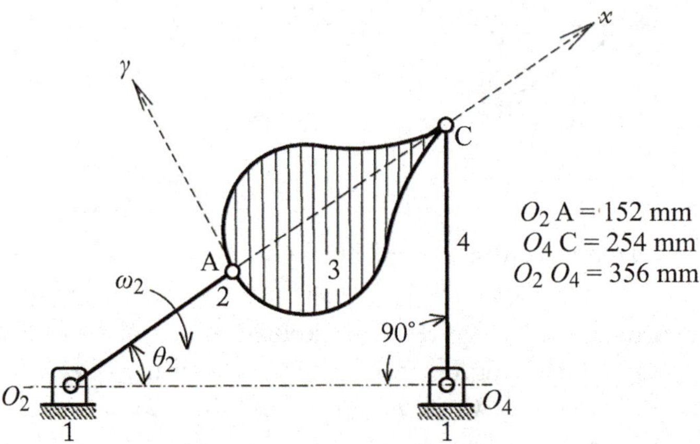

Figure 5.14: A four-bar mechanism.

5.4. For the linkage shown in Figure 5.15, link 2 rotates ccw at a constant angular velocity, $\omega_2 = 2.0$ rad/s. By applying the unit vectors method, determine the velocity $\vec{v}_B$ and acceleration $\vec{a}_B$. Note that the axes of the *RFR* with its origin at A and attached to link 3 are indicated in the figure in which dimensions are in mm.

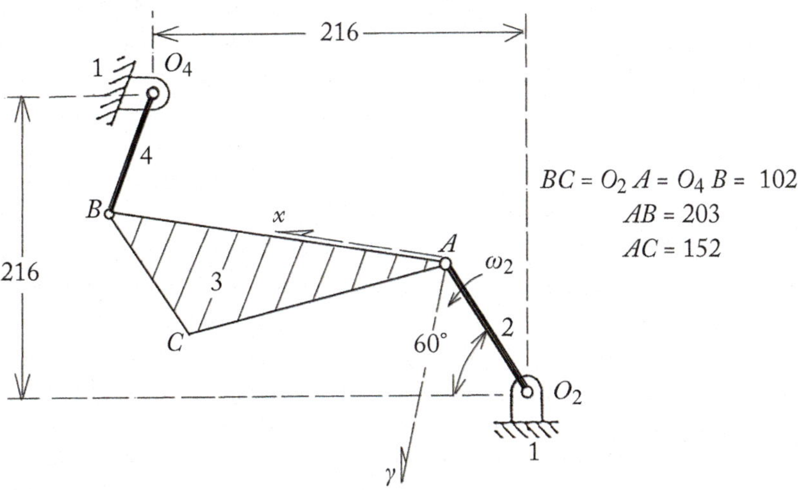

Figure 5.15: Four-bar mechanism with RFR.

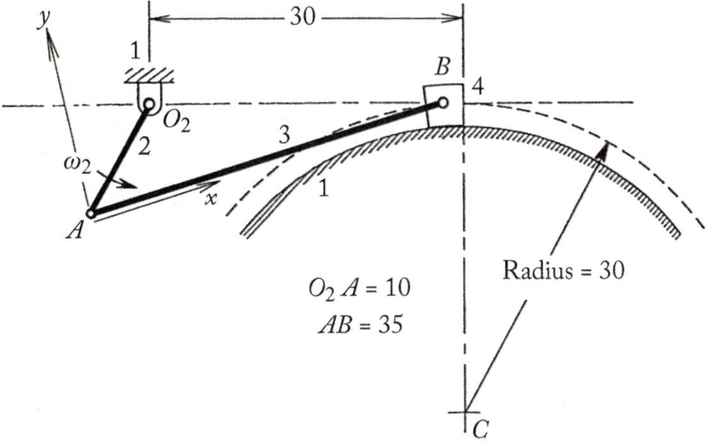

Figure 5.16: Four-bar mechanism with a slider on circular ground.

5.5. For the linkage shown in Figure 5.16, link 2 rotates ccw at a constant angular velocity, $\omega_2 = 2.0$ rad/s and $\angle AO_2 B = 120°$. By applying the unit vectors method, determine the velocity $\vec{v}_B$ and acceleration $\vec{a}_B$. Note that the axes of the *RFR* with its origin at A and attached to link 3 are indicated in the figure in which dimensions are in mm.

5.6. The linkage shown in Figure 5.17 is known as Geneva mechanism. The origin of the *RFR* is at A and it is attached to link 3. The angular velocity of link 3 is $\omega_3 = 10$ rad/s, ccw, and its angular acceleration $\alpha_3 = 100$ rad/s^2, ccw. Given angles are $\angle AO_3O_2 = 40.5°$ and $\angle O_3O_2A = 25.1°$. By using the method of unit vectors, determine the velocity $\vec{v}_{A_2}$ and acceleration $\vec{a}_{A_2}$.

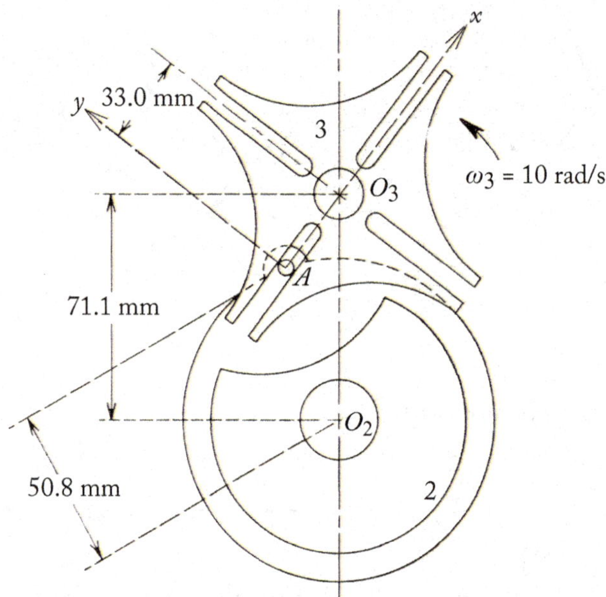

Figure 5.17: Geneva mechanism.

REFERENCES

[1] Bolotin, V. V., *The Dynamic Stability of Elastic Systems*, San Francisco, Holden-Day, Inc., 1964. 53

[2] Ince, E. L., On a general solution of Hill's equation, *Monthly Notices Royal Astronomical Society*, 75, pp. 436-448, 1915. 53

[3] Mabie, H. H. and Ocvirk, F. W., *Mechanisms and Dynamics of Machinery*, 3rd edition-SI Version, New York, John Wiley & Sons, 1978. 63

CHAPTER 6

Graphical Approaches in Motion Studies

In addition to the analytical approaches presented in Chapter 5, there are graphical approaches employed in the studies of motion in machinery. In this chapter two commonly applied methods are introduced. They are the method of instantaneous centers, and the image method. These two methods are presented, respectively, in Sections 6.1 and 6.2.

6.1   INSTANTANEOUS CENTER METHOD

This method is also known as the method of instantaneous center of rotation. It consists of two stages. The first stage is the determination of instantaneous or instant centers. The second stage is the application of the instant centers for the evaluation of velocities. These two stages will be included in the following sections. It may be appropriate to point out that frequently only the relevant instant centers are required in the computation of velocities.

6.1.1   DETERMINATION OF INSTANT CENTERS (OF ROTATION)

Before the determination of instant centers (IC), the definition, notation, number of IC in a planar mechanism, and Kennedy's theorem are introduced in the following.

(a) *Definition*

- An instantaneous or instant center (IC) is a point at which the two bodies have no relative velocity.
- An IC is a point about which one body may be considered to rotate relative to the other body at a given instant.
- An IC is a point in both bodies.

(b) *Notation*

Various notations for IC have been used in the literature. Examples are:

$$I_{ij} \text{ or } I_{i,j} \quad \text{for example} \quad I_{14} \text{ or } I_{1,4}$$
$$ij \text{ or } i,j \quad \text{for example} \quad 14 \text{ or } 1,4$$

In this chapter the notation ij is adopted unless it is stated otherwise.

(c) *Number of Instant centers N_{IC}*

If a linkage consists of n links, including the ground link, then the number of IC is given by

$$N_{IC} = \frac{n(n-1)}{2}. \tag{6.1}$$

Some examples are provided in Table 6.1 in the following.

Table 6.1: Examples

n	N_{IC}
2	1
3	3
4	6
5	10
6	15
7	21

In general, some of the IC are obvious while the remaining non-obvious IC have to be determined by applying Kennedy's rule and "circle diagram" which will be introduced later in an example.

(d) *Kennedy's rule (theorem)*

The Kennedy's rule or theorem states that for any three bodies in planar motion there are three IC and the latter must lie on the same straight line, as shown in Figure 6.1.

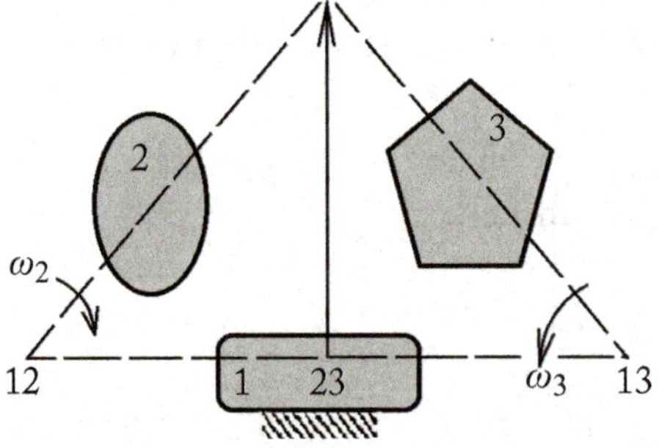

Figure 6.1: Instant centers of three rigid bodies.

Suppose in Figure 6.2 the velocity at point A is $\vec{v}_A$ then it must be perpendicular to line connecting point A and IC 12 or perpendicular to the line connecting point A and IC

13. As the two velocities at the point are not in the same direction, point A cannot be an IC. In other words, for point A to be an IC there is one unique velocity which must be perpendicular to the straight line connecting all three IC.

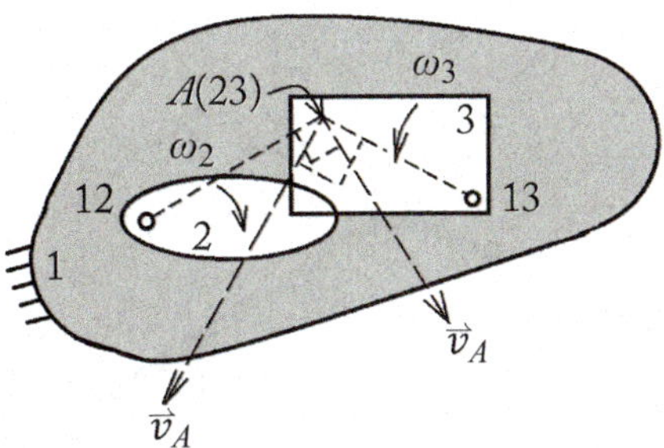

Figure 6.2: Three bodies used in proof of Kennedy's rule.

(e) *Determination of instantaneous centers*

- An IC may be located beyond the physical dimensions of a link.
- All pin-joints are IC for at such points, the two connected links have the same velocity.
- A straight slider will have an IC that is at infinity and along the direction normal to the path of the slider. A corollary to this fact is that a circular slider will have an IC at the center of curvature.
- For those IC that are not obvious, Kennedy's rule can be applied together with the "circle diagram." This will be illustrated in the following simple example.

Example 6.1
A slider mechanism is shown in Figure 6.3. Note that the configuration diagram of the mechanism in the latter figure has been drawn to scale. Locate all the instant centers in the mechanism.

Solution:
The solution consists of the following two stages. The first stage is the determination of the number of instant centers (IC), and the second is the location of IC. They are outlined in the following.

(a) *Determine the number of instant centers*

According to Equation (6.1), since $n = 4$, hence the number of IC is 6.

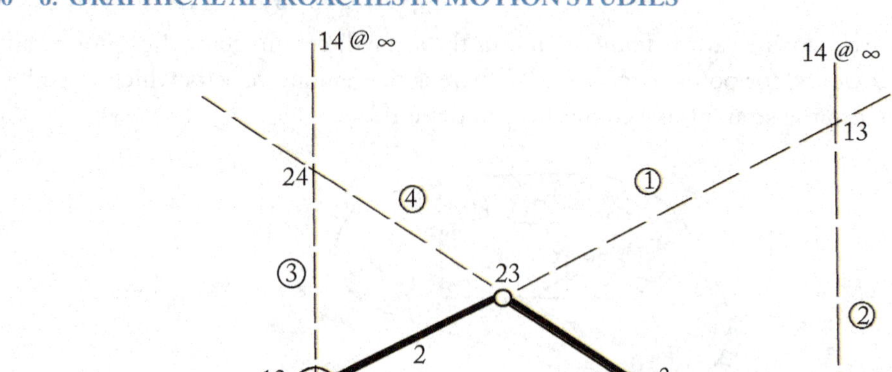

Figure 6.3: Slider mechanism.

(b) *Location of IC*

This is divided into two parts: the locations of obvious IC and non-obvious ones.

Obvious IC

From Figure 6.3 the obvious IC are: 12, 23, 34, and 14.

Non-obvious IC

Since $N_{IC} = 6$ therefore the number non-obvious IC is $6 - 4 = 2$. These two non-obvious IC are determined by using the "circle diagram" and Kennedy's rule as shown in the following five steps.

Step 1: A "circle diagram" is drawn and a polygon with the number of sides equal to the number of links in the mechanism is constructed. Note that for identification and ease of tracking purpose the obvious IC are represented by full lines while the non-obvious IC are denoted by dash lines. The two integers (at the vertices of the polygon) connected by every side of the polygon constitute the IC.

Step 2: Locate every non-obvious IC which is denoted by the dash line in the "circle diagram" by selecting a pair of triangles such that a common side is represented by the dash line. In each selected triangle there are two full lines representing the two determined or known IC.

Step 3: Construct a straight line that is collinear with the two obvious IC of each triangle in the last step, **Step 2**. A circled integer is placed next to the constructed line in the configuration diagram for identification purpose. Repeat this step for the other triangle. The intersection point of the pair of constructed lines is the required IC.

Step 4: Once the non-obvious IC is located superimpose the dash line by a full line, meaning now the non-obvious IC has been located.

Step 5: Repeat **Steps 3** and **4** until all non-obvious IC are located.

For the present example, there are two non-obvious IC and there are 4 links with 4 obvious IC so that the "circle diagram" is shown below (Figure 6.4). Note that the two dash lines or non-obvious IC are the two diagonals in the "circle diagram" on the lhs below while the non-obvious IC and the two triangles as well as their associated IC are listed on the rhs. On the rhs of this "circle diagram" the circled integers are the constructed lines in Figure 6.3. In the present example, either one of the two non-obvious IC can be located first. Thus, one can proceed with non-obvious IC 13, where the thick arrow denotes the line joining the two IC being identified by the circled integer in Figure 6.3.

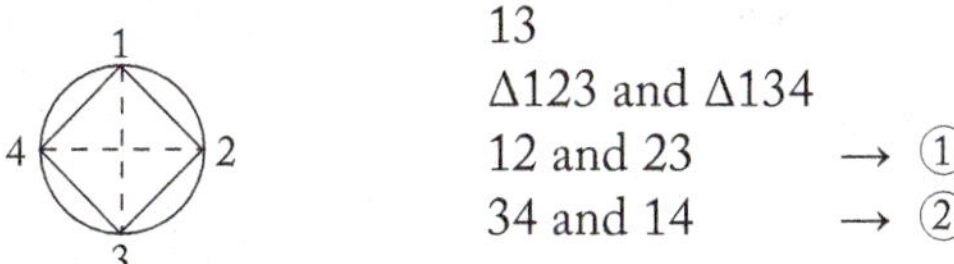

Figure 6.4: Location of instant center 13.

Similarly, by repeating the foregoing steps, the remaining non-obvious IC 24 can be located as in the following (Figure 6.5):

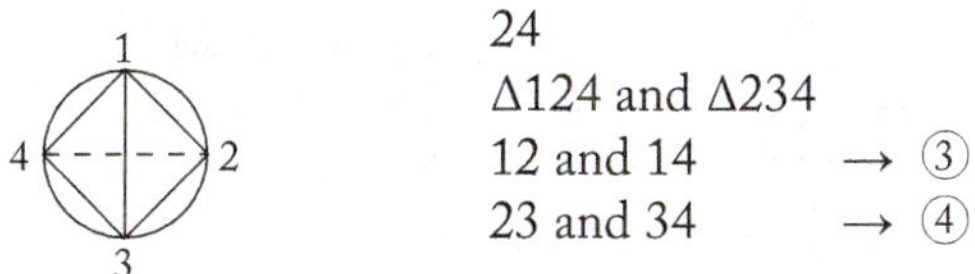

Figure 6.5: Location of instant center 24.

6.1.2 APPLICATION OF INSTANT CENTERS

Note that while the IC method is very useful, in the sense that it can provide a complete picture of the velocities in the linkage, it cannot be applied to the determination of accelerations and angular accelerations.

Consider the four-bar linkage in Figure 6.6 below. The angular velocity of link 2, ω_2 is given and its direction is shown in the figure. It is assumed that this figure is constructed in scale so that in the determination of the obvious and non-obvious IC one can make use of this figure. The objectives are to find the unknown velocities. The application of the IC method for the determination of velocities consist of the following six steps.

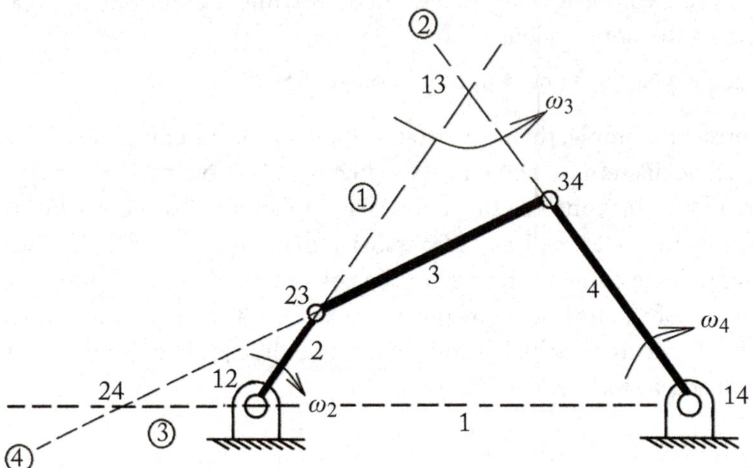

Figure 6.6: A four-bar mechanism with all instant centers.

Step 1: $N_{IC} = n(n-1)/2 = 4(4-1)/2 = 6.$

  Obvious IC: 12, 14, 23, 34.

Step 2: Non-obvious IC (Figure 6.7):

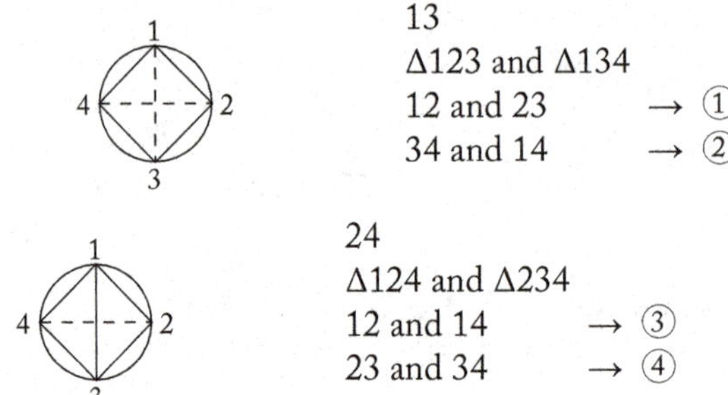

Figure 6.7: Location of non-obvious instant centers.

Step 3: Instant center 13 is a point in link 1, the ground link, and in link 3, the connecting rod. Link 3 rotates about IC 13 with ω_3.

Step 4: Since link 3 rotates about IC 13, as shown in Figure 6.6, velocities of points on link 3 are proportional to their distances from 13. For example, one may state the ratio of velocities of points 34 and 23 as

$$\frac{v_{34}}{v_{23}} = \frac{13 \to 34}{13 \to 23} \tag{6.2}$$

where $13 \to 34$ and $13 \to 23$ are the distances between IC 13 and 34, and IC 13 and 23, respectively, in the scaled linkage drawing in Figure 6.6.

The magnitude of velocity $\vec{v}_{23}$ is given by the product of the length of link 2 and angular velocity ω_2. Applying the magnitude of $\vec{v}_{23}$, the magnitude of $\vec{v}_{34}$ may be determined by Equation (6.2).

A word about the notation is in order. In the application of the IC method the two subscripts or indices in the velocity terms denote the velocity at that particular IC and they should not be confused with relative velocities at the points representing by the two subscripts.

Step 5: The angular velocity ω_3 is given by

$$\omega_3 = \frac{v_{23}}{13 \to 23}. \tag{6.3}$$

Note that to provide the correct answer the chosen scales for the velocity and configuration diagrams have to be applied.

Step 6: The angular velocity of link 4 follows immediately. Since

$$v_{24} = \omega_2(12 \to 24), \qquad v_{24} = \omega_4(14 \to 24).$$

Equating the above two expressions, one obtains

$$\omega_4(14 \to 24) = \omega_2(12 \to 24)$$

or

$$\frac{\omega_4}{\omega_2} = \frac{(12 \to 24)}{(14 \to 24)}. \tag{6.4}$$

Example 6.2

With the scaled configuration diagram of the mechanism shown in Figure 6.8, locate all instant centers. Determine the velocity of slider 5 using the method of instant centers. Given that $\omega_2 = 100$ rad/s ccw.

Solution:

(a) *Construct a scaled diagram and calculate number of IC*

Since Figure 6.8 is scaled and therefore the location of all IC can be performed using this figure.

Since $n = 5$, hence the number of IC is $N_{IC} = n(n-1)/2 = 10$.

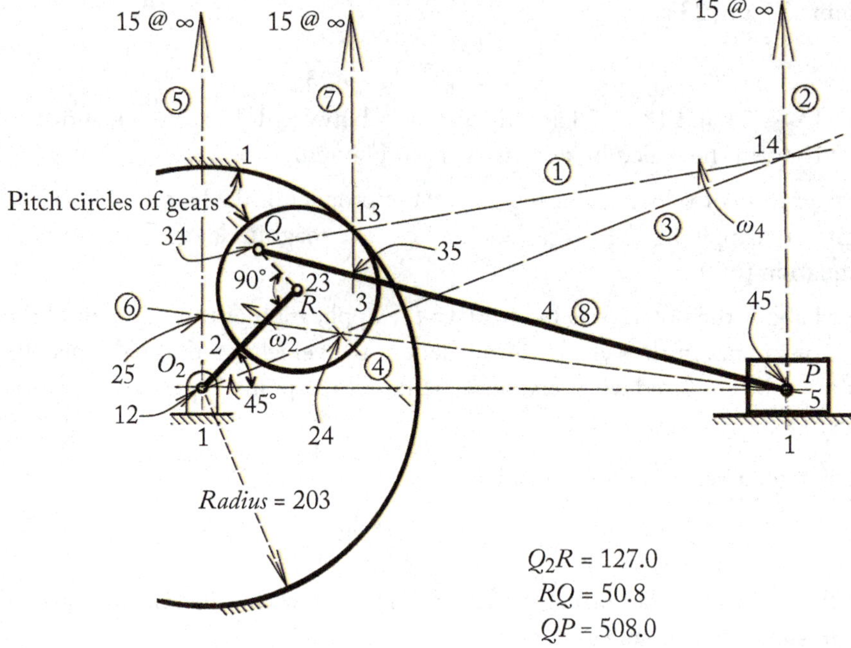

Figure 6.8: Five-bar mechanism (dimensions in mm).

(b) **Obvious and non-obvious IC**

From Figure 6.8, the obvious IC are: 12, 23, 13, 34, 45, and 15.

There are four non-obvious IC since $N_{IC} = 10$ and the number of obvious IC is six. Thus, these non-obvious IC are located as in the following (Figure 6.9).

(c) **Velocity solution**

Starting from the known magnitude and direction of velocity at point R or IC 23 in Figure 6.8 since the angular velocity ω_2 and length of link 2 are given,

$$v_{23} = \omega_2(12 \rightarrow 23) = 5\omega_2 \qquad \text{units/s}$$

since $12 \rightarrow 23 = 5$ units from Figure 6.8. Note that for the present problem 1 unit $=$ 25.4 mm. Also,

$$v_{23} = \omega_3(13 \rightarrow 23) = 3\omega_2 \qquad \text{units/s}$$

since

$$(13 \rightarrow 23) = 3 \quad \text{units.}$$

Thus,

$$\omega_3 = \frac{5}{3}\omega_2 \qquad \text{rad/s}$$

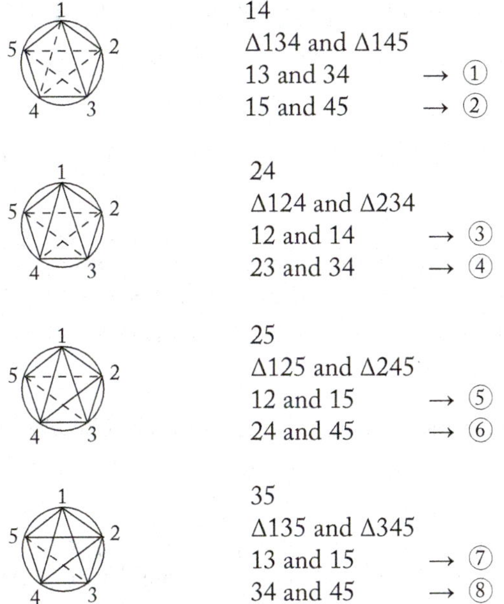

Figure 6.9: Circle diagrams for non-obvious instant centers.

which enables one to write

$$v_{34} = \omega_3(13 \to 34) = 3.5\omega_3, \quad \text{since} \quad 13 \to 34 = 3.5 \text{ units.}$$

Hence,

$$v_{34} = \frac{5}{3}\left(\frac{7}{2}\right)\omega_2 = 5.83\omega_2.$$

Furthermore,

$$v_{34} = \omega_4(14 \to 34), \quad \text{but} \quad 14 \to 34 = 19.7 \text{ units}$$

$$\omega_4 = \frac{5}{3}\left(\frac{7}{2}\right)\frac{\omega_2}{19.7} = 0.296\omega_2.$$

Now,

$$v_{45} = \omega_4(14 \to 45), \quad \text{but} \quad 14 \to 45 = 8.5 \text{ units.}$$

Therefore,

$$v_{45} = \omega_4(8.5) = 2.516\omega_2 = 2.516(100) \text{ unit/s.}$$

The required velocity

$$\vec{v}_P = \vec{v}_{45} = 251.6 \times 25.4 \text{ mm/s} \leftarrow = 6.3906 \text{ m/s} \leftarrow.$$

That is, the velocity at P, $\vec{v}_P$ is 6.3906 m/s directed horinzontally toward point O_2.

6.2   IMAGE METHOD FOR MOTION STUDIES

This method is based on the principle of similar triangles or proportionality. The main advantages are:

- It gives a complete picture of the velocity distribution and acceleration distribution in the mechanism.

- It is good for quick checking with analytical solution for low speed problems but not good for high speed linkages due to significant error occurred in the graphical construction. In order to improve the accuracy, one may construct very large velocity and acceleration diagrams.

  The steps in the image method are illustrated in the following two examples.

Example 6.3
The constant angular velocity of the crank in Figure 6.10 is given as ω_2 and direction as shown. The mechanism has been constructed with a chosen scale. By graphical construction, determine

(a) the angular velocity, and

(b) the angular acceleration of the disk, link 4, which rolls, without slipping, on ground 1.

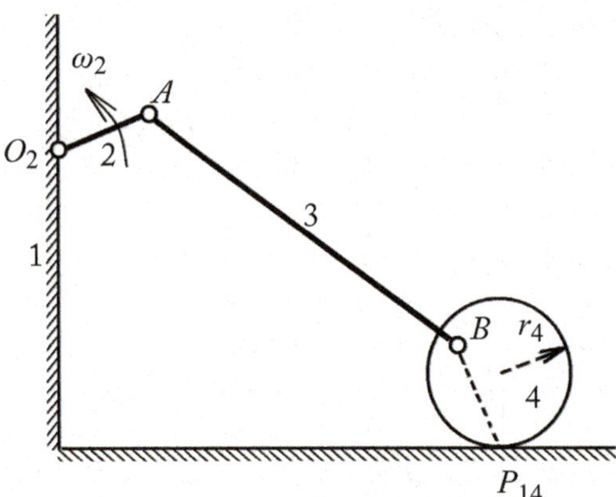

Figure 6.10: Mechanism with a circular disk link.

Solution:
The solutions are presented in the following. Note that all dimensions are in Figure 6.10 since a scale has been assumed.

(a) *Determination of $\vec{\omega}_4$*

Step 1: Determine input velocity $\vec{v}_A = oa$ which is to be constructed in Figure 6.11 since $\vec{v}_A = \vec{\omega}_2 \times \vec{O_2 A}$. That is, its magnitude $v_A = \omega_2\,(O_2 A)$ is known. The velocity $\vec{v}_A$ is known and perpendicular to line $O_2 A$, and its specific direction is determined by the sense of $\vec{\omega}_2$. It may be appropriated to point out that the starting or reference point o is chosen arbitrarily in the space for construction.

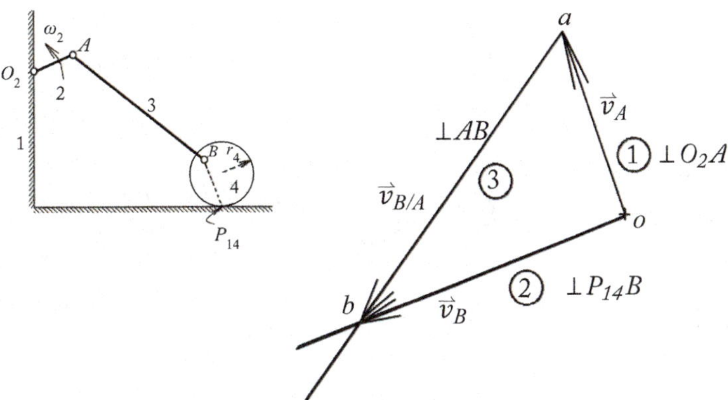

Figure 6.11: Graphical construction of velocity polygon.

Step 2: Construct $\vec{v}_B = ob$, based on $\vec{v}_B = \vec{v}_A + \vec{v}_{B/A}$. Point b is located at the intersection of two lines ob, drawn perpendicular to $P_{14}B$, and ab perpendicular to AB.

Choose a scale so that the following velocity diagram in Figure 6.11 can be constructed in accordance with the above steps. Note that in general this scale is different from that for the geometrical configuration diagram in Figure 6.10.

Step 3: *Calculation of $\vec{\omega}_4$*

With reference to Figure 6.11,

$$\omega_4 = \frac{v_B}{P_{14}B} = \frac{ob}{P_{14}B}, \qquad \text{ccw.}$$

It is important to emphasize that the scales for the velocity and configuration diagrams are different thus, care has to be taken in order to obtain the correct value of ω_4.

(b) *Determination of $\vec{\alpha}_4$*

Choose a scale so that the acceleration diagram in Figure 6.12 can be constructed.

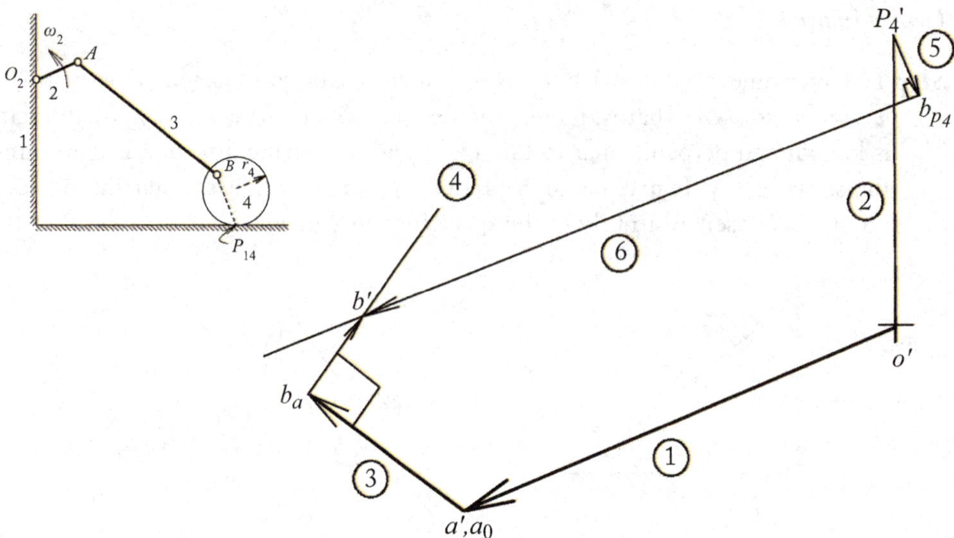

Figure 6.12: Graphical construction of acceleration polygon.

Step 1: Determine input acceleration $\vec{a}_A = o'a'$ since $a_A = \frac{(oa)^2}{O_2 A}$ and its direction is from A to O_2 (recall, the normal component of the acceleration at a point is always directed from that point toward the center of curvature). Note that only the normal component of the acceleration at point A is evaluated since the tangential component is zero due to $\vec{\alpha}_2 = 0$ for ω_2 is constant. Note also that in the acceleration diagram primes are placed next to the alphabets to distinguish them from those used in the velocity construction. That is, for example, $o'a'$ is different from oa.

Step 2: Determine $\vec{a}_{P_4} = o'p'_4$ since $a_{P_4} = \omega_4^2 r_4$. Note that $r_1 = \infty$ and r_4 is the radius of the disk. The direction is away from link 1 (ground link). That is, direction $\uparrow$. In the foregoing $\vec{a}_{P_4}$ is $\vec{a}_{P_{14}}$.

- Note that this step is made necessary by the fact that $\vec{a}_B$ cannot be computed directly from the vector equation $\vec{a}_B = \vec{a}_A + \vec{a}_{B/A}$ because the radius of curvature of the path of point B is not known.

Step 3: Construct $\vec{a}_B = o'b'$ based on $\vec{a}_A + \vec{a}_{B/A} = \vec{a}_{P_4} + \vec{a}_{B/P_4}$. Point b', the end point of the line representing $\vec{a}_B$, is located at the intersection of $(\vec{a}_{B/A})_t = b_a b'$ with $(\vec{a}_{B/P_4})_t = b_{P_4} b'$

- The computations provided in **Steps 1–3** above are summarized in the following table, Table 6.2. It is good practice to include all the computation in a single table. This is particularly so if there are many links in a given mechanism. It will always enable one to check if any quantity is missed or at the initial stage a particular quantity is unable to be found. In this way, after the identified quantities are provided one can return to complete the particular quantity that has been missed.

Table 6.2: Distribution of acceleration quantities of given linkage

Term	Normal Acceleration			Tangential Acceleration		
	Magnitude	Direction	Vector	Magnitude	Direction	Vector
$\vec{a}_A = o'a'$	$(oa)^2/(O_2 A)$	$A \rightarrow O_2$	$o'a_o$	$0,\ \alpha_2 = 0$	none	$a_o a'$
$\vec{a}_{P_4} = o'p_4'$	$\omega_4^2\, r_4$	$\uparrow$		$0,\ \because$	no	slip
$\vec{a}_{B/A} = a'b'$	$(ab)^2/(AB)$	$B \rightarrow A$	$a'b_a$	unknown	$\|\vec{v}_{B/A}$	$b_a b'$
$\vec{a}_{B/P_4} = bp_4 b'$	$\omega_4^2(P_4 B)$	$B \rightarrow P_4$	$p_4' b p_4$	unknown	$\|\vec{v}_B$	$bp_4 b'$

Step 4: *Calculation of $\vec{\alpha}_4$*

From Figure 6.12, the angular acceleration of link 4,

$$\alpha_4 = \left(a_{B/p_4}\right)_t / (P_4 B) = b_{p_4} b' / (P_4 B), \qquad \text{ccw.}$$

Note that the scales of the numerator and denominator terms are different and therefore care has to be taken to convert these quantities into their actual values.

Summary of steps in acceleration graphical construction
In Figure 6.12, the circled integers denote the steps in the construction. These steps should not be confused with those in the computation. Thus, the sequences are as follows.

① $(\vec{a}_A)_n$, magnitude known, $\mathbf{A} \rightarrow \mathbf{O}_2$.

② $\left(\vec{a}_{P_4}\right)_n$, magnitude known, direction upward.

③ $\left(\vec{a}_{B/A}\right)_n$, magnitude known, $\mathbf{B} \rightarrow \mathbf{A}$.

④ $\left(\vec{a}_{B/A}\right)_t$, magnitude unknown, $\|$ to $\vec{v}_{B/A}$ or $\perp \left(\vec{a}_{B/A}\right)_n$.

⑤ $\left(\vec{a}_{B/P_4}\right)_n$, magnitude known, $\|$ to $\mathbf{B} \rightarrow \mathbf{P}_4$.

⑥ $\left(\vec{a}_{B/P_4}\right)_t$, magnitude unknown, $\|$ to $\vec{v}_B$.

Example 6.4
The angular velocity $\vec{\omega}_2$ and angular acceleration $\vec{\alpha}_2$ of the crank, shown in Figure 6.13, are included. In addition, the dimensions of the links are given such that the configuration diagram in Figure 6.13 has been drawn to scale. Determine the:

(a) velocity at point D,

(b) angular velocities of links 3, 4, and 5,

(c) acceleration at point D, and

(d) angular accelerations of links 3, 4 and 5.

Solutions:
All dimensions in Figure 6.13 are given and the latter figure has been drawn to scale. One can construct the velocity diagram as in the following.

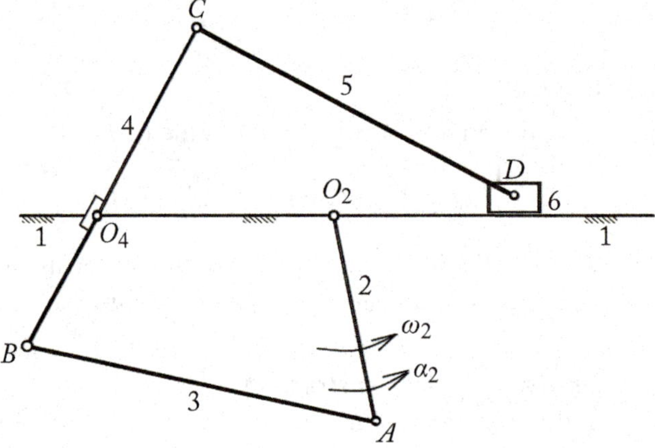

Figure 6.13: Scaled diagram of a six-bar mechanism.

(a) **Determination of $\vec{v}_D$**

Choose a scale for the velocity diagram, shown in Figure 6.14 which is constructed by applying the following steps.

Step 1: Determine input velocity $\vec{v}_A = oa$, since $\vec{v}_A = \vec{\omega}_2 \times \vec{O_2 A}$. That is, its magnitude $v_A = \omega_2 (O_2 A)$. $\vec{v}_A$ is perpendicular to line $O_2 A$, and its specific direction is determined by the sense of ω_2.

Step 2: Construct $\vec{v}_B = ob$, based on $\vec{a}_B = \vec{a}_A + \vec{a}_{B/A}$. Point b is located at the intersection of two lines ob, drawn perpendicular to $O_4 B$, and ab perpendicular to AB.

Step 3: Draw the velocity at point C, $\vec{v}_C = oc$, by the image method. Since C, O_4, and B are collinear points of link 4, the line cob is similar to $CO_4 B$. Thus, the velocity at C is given by $oc = (ob/O_4 B) O_4 C$.

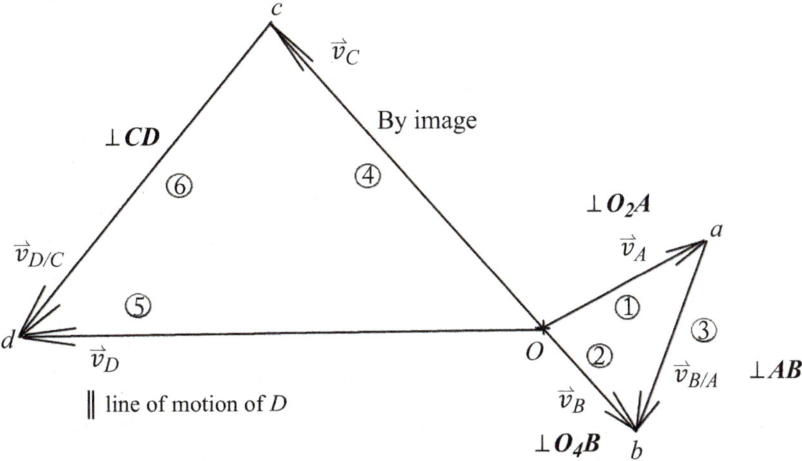

Figure 6.14: Graphical construction of velocity polygon.

Step 4: Draw the velocity at point D, $\vec{v}_D = od$, based on $\vec{v}_D = \vec{v}_C + \vec{v}_{D/C}$. Point d is located at the intersection of two lines od, drawn parallel to the line of motion of point D (that is, along the horizontal direction) and cd, drawn perpendicular to CD (recall, relative velocity cd is perpendicular to the relative position vector CD in the configuration diagram).

Thus, from the above velocity diagram the velocity at point D, od is obtained.

(b) ***Determination of $\vec{\omega}_3$, $\vec{\omega}_4$, and $\vec{\omega}_5$***

With reference to the above constructed velocity diagram, the required angular velocities are determined as follows:

$$\omega_3 = ab/(AB), \quad \text{ccw}; \quad \omega_4 = oc/(O_4C), \quad \text{ccw}; \quad \text{and} \quad \omega_5 = cd/(CD), \quad \text{cw}.$$

Note that the numerator and denominator terms have different scales.

(c) ***Determination of $\vec{a}_D$***

Choose a scale for the acceleration diagram that is presented in Figure 6.15.

Step 1: Calculate input acceleration $\vec{a}_A = o'a'$. The magnitude of the normal component $(a_A)_n = \frac{(oa)^2}{O_2A}$ and its direction is from A to O_2. Note that the tangential component is not zero and is represented by the line $(\vec{a}_A)_t = a_oa'$.

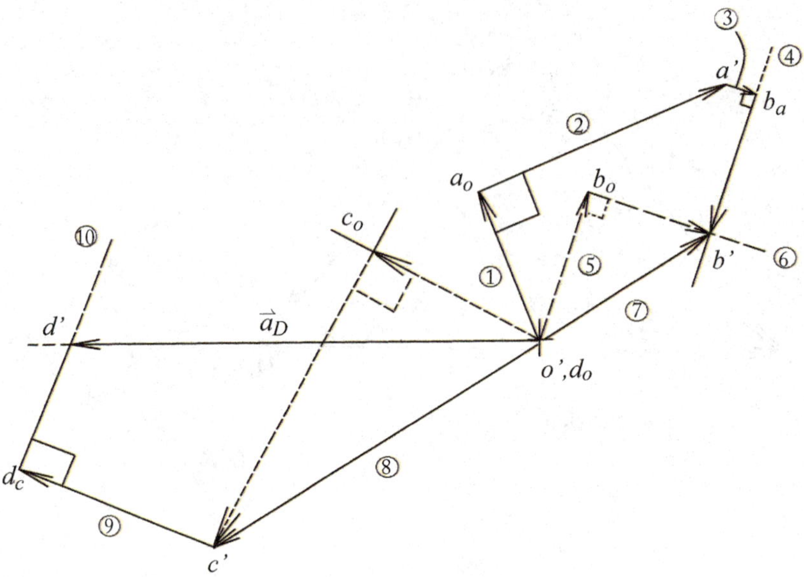

Figure 6.15: Graphical construction of acceleration polygon.

Step 2: Construction of $\vec{a}_B = o'b'$ based on $\vec{a}_B = \vec{a}_A + \vec{a}_{B/A}$. Point b', the end point of the line representing $\vec{a}_B$, is located at the intersection of $(\vec{a}_B)_t = b_o b'$ with $(\vec{a}_{B/A})_t = b_a b'$.

Step 3: Construction of $\vec{a}_{B/A} = a'b'$ with magnitude of normal component $(a_{B/A})_n = \frac{(ab)^2}{AB}$ and its direction is from B toward A. Thus, $(\vec{a}_{B/A})_n = a'b_a$. Magnitude of $(\vec{a}_{B/A})_t$ is unknown but its direction is parallel to $\vec{v}_{B/A}$ or perpendicular to $(\vec{a}_{B/A})_n = a'b_a$. Thus, $(\vec{a}_{B/A})_t = b_a b'$.

Step 4: Construction of $\vec{a}_C = o'c'$, based on the image method. That is, $c'o'b'$ is constructed similar to CO_4B in the configuration diagram.

Step 5: Construction of $\vec{a}_D = o'd'$ based on $\vec{a}_D = \vec{a}_C + \vec{a}_{D/C}$. Point d', the end point of the line representing $\vec{a}_D$, is located at the intersection of $(\vec{a}_D)_t = d_o d'$ with $(\vec{a}_{D/C})_t = d_c d'$.

- The calculations for **Steps 1–5** are summarized in Table 6.3.

Table 6.3: Distribution of acceleration quantities of given linkage

Term	Normal Acceleration			Tangential Acceleration		
	Magnitude	Direction	Vector	Magnitude	Direction	Vector
$\vec{a}_A = o'a'$	$(oa)^2/(O_2A)$	$A \to O_2$	$o'a_o$	$\alpha_2(O_2A)$	$\| \vec{v_A}$	a_oa'
$\vec{a}_B = o'b'$	$(ob)^2/(O_4B)$	$B \to O_4$	$o'b_o$	unknown	$\| \vec{v_B}$	b_ob'
$\vec{a}_{B/A} = a'b'$	$(ab)^2/(AB)$	$B \to A$	$a'b_a$	unknown	$\| \vec{v}_{B/A}$	b_ab'
$\vec{a}_D = o'd'$	$0, \because \rho_D = \infty$	none	$o'd_o$	unknown	$\| \vec{v_D}$	$d_o \to d'$
$\vec{a}_c = o'c'$	by image	method,	$c'o'b'$	$\propto$	CO_4B	
$\vec{a}_{D/C} = c'd'$	$(cd)^2/(CD)$	$D \to C$	$c'd_c$	unknown	$\| \vec{v}_{D/C}$	d_cd'

Summary of steps in acceleration graphical construction

① $\left(\vec{a}_A\right)_n$, magnitude known, $\mathbf{A} \to \mathbf{O}_2$.

② $\left(\vec{a}_A\right)_t$, magnitude known, $\|$ to $\vec{v}_A$.

③ $\left(\vec{a}_{B/A}\right)_n$, magnitude known, $\mathbf{B} \to \mathbf{A}$.

④ $\left(\vec{a}_{B/A}\right)_t$, magnitude unknown, $\|$ to $\vec{v}_{B/A}$ or $\perp \left(\vec{a}_{B/A}\right)_n$.

⑤ $\left(\vec{a}_B\right)_n$, magnitude known, $\mathbf{B} \to \mathbf{O}_4$.

⑥ $\left(\vec{a}_B\right)_t$, magnitude unknown, $\|$ to $\vec{v}_B$.

⑦ $\vec{a}_B$, connecting points o' and b'.

⑧ $\vec{a}_C = o'c'$, by image method. Since $\vec{a}_B = o'b'$ is known from the last two steps one can use the relation, $\frac{o'c'}{o'b'} = \frac{O_4C}{O_4B}$ to obtain $o'c'$ which is parallel to $o'b'$.

⑨ $\left(\vec{a}_{D/C}\right)_n$, magnitude known, $\mathbf{D} \to \mathbf{C}$.

⑩ $\left(\vec{a}_{D/C}\right)_t$, magnitude unknown, $\|$ to $\vec{v}_{D/C}$ or $\perp \left(\vec{a}_{D/C}\right)_n$.

(d) *Calculation of angular accelerations*

With reference to Figure 6.15, angular acceleration of link 3,

$$\alpha_3 = \left(a_{B/A}\right)_t /(AB) = b_ab'/(AB), \qquad \text{ccw.}$$

Similarly, angular acceleration of link 4,

$$\alpha_4 = \frac{(a_C)_t}{O_4C} = \frac{o'c_o}{O_4C} \qquad \text{ccw.}$$

Note that $o'c_o$ from the acceleration diagram is used and not c_oc' since $o'c'$ is the vector sum of the normal and the tangential components. The normal component of $\vec{a}_C$ is in the direction from C to O_4. That is, it is parallel to link 4. The tangential component, as indicated in Figure 6.15, is perpendicular to link 4. Recall that to evaluate α_4 one has to use the tangential component.

Finally, angular acceleration of link 5,

$$\alpha_5 = \left(a_{D/C}\right)_t / (CD) = d_c d' / (CD), \qquad \text{ccw.}$$

- Note that in the above calculations the lengths from the acceleration diagram, for example, $b_a b'$, $o' c_o$, and $d_c d'$ have to be converted back to their actual values by using the scale chosen for the construction of the acceleration diagram.

6.3   EXERCISES

6.1.   A mechanism having six links is shown in Figure 6.16 in which the dimensions are in cm. Construct a scaled configuration diagram. Applying the method of instant centers, find the velocity at point B of the slider (link 4), and angular velocities of links 3, and 5 if the velocity at point D (link 6) is $\vec{v}_D = 4.57 \frac{m}{s}$ upward as indicated in the figure.

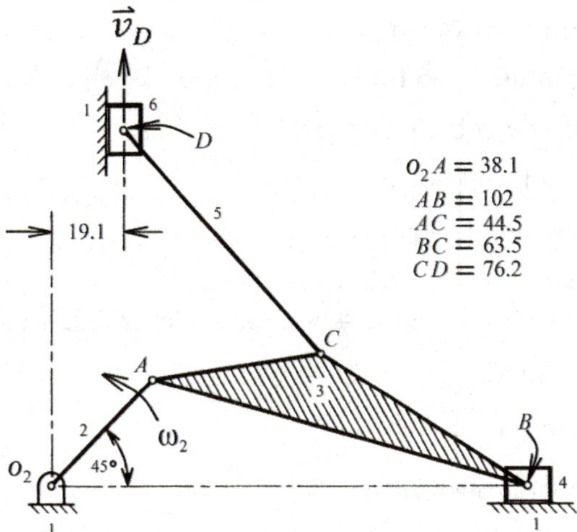

Figure 6.16: Six-bar mechanism (dimensions are in cm).

6.2.   For the mechanism shown in Figure 6.17, by applying the method of instant centers, determine the angular velocity of link 5 if the angular velocity of crank 2 is $\vec{\omega}_2 = 1000$ rpm clockwise. Find also the velocity at point D.

6.3.   In the mechanism shown in Figure 6.18 the length of link 2 is 200 mm and the angular velocity of link is 1000 rpm ccw. Assume the figure has been drawn in scale. Applying

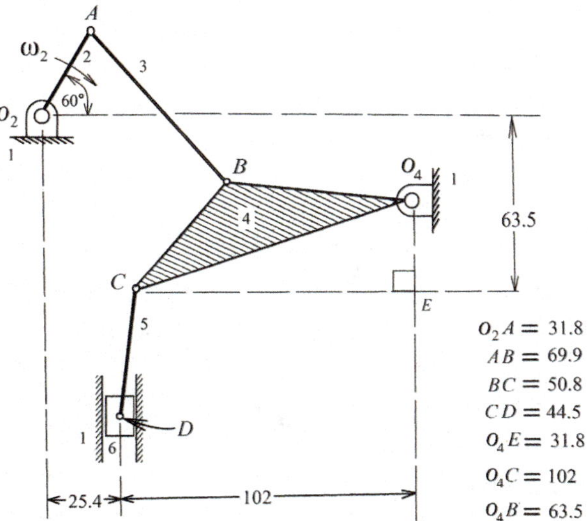

Figure 6.17: Six-bar mechanism (dimensions are in cm).

the instant center method, determine the velocities at points C and D, and the angular velocities of links 3, 4, and 5.

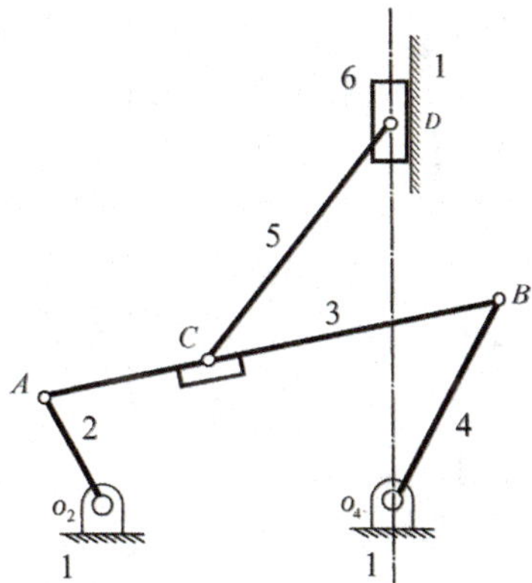

Figure 6.18: Scaled diagram of a six-bar mechanism with a slider.

6.4. In the mechanism shown in Figure 6.19, link 2 rotates at a constant angular velocity, $\omega_2 = 200$ rad/s ccw. Construct graphically the velocity and acceleration polygons, and determine angular velocity and angular acceleration of link 4.

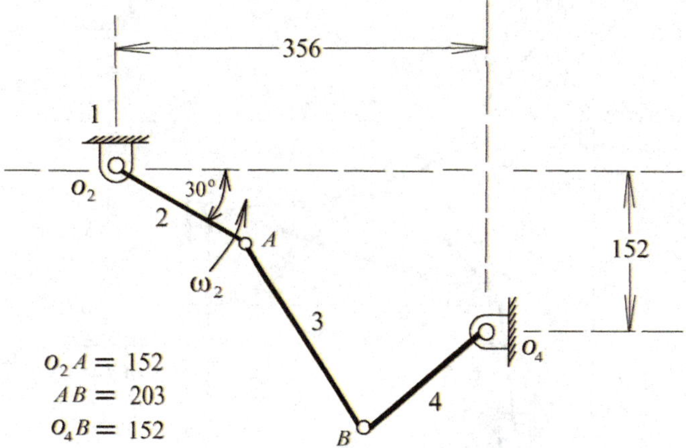

Figure 6.19: A four-bar mechanism.

6.5. A toggle mechanism is shown in Figure 6.20 with link 2 rotating at a constant angular velocity, $\omega_2 = 20$ rad/s ccw. Construct the velocity and acceleration polygons. Determine $\vec{v}_c$ and $\vec{a}_c$ of the slider (link 6).

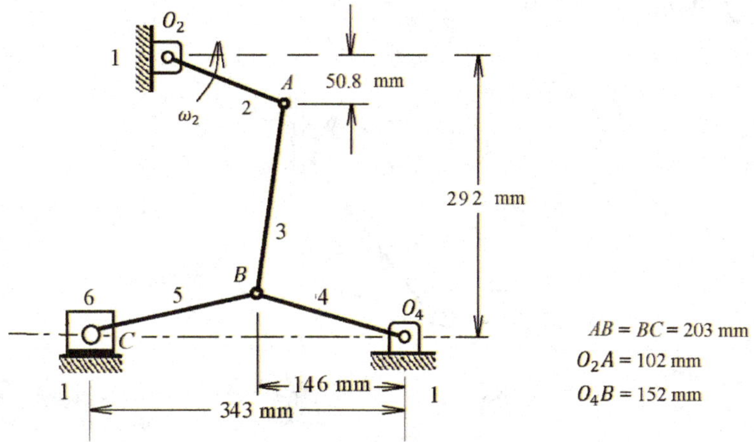

Figure 6.20: Toggle mechanism.

6.6. A double slider-crank mechanism is shown in Figure 6.21. The crank (link 2) rotates at a constant angular velocity of $\omega_2 = 20$ rad/s ccw. Construct the velocity and acceleration polygons. With the constructed velocity polygon, determine the velocities at points B

and C, $\vec{v}_B$ and $\vec{v}_C$. In addition, with the constructed acceleration polygon, determine $\vec{a}_B$ and $\vec{a}_C$.

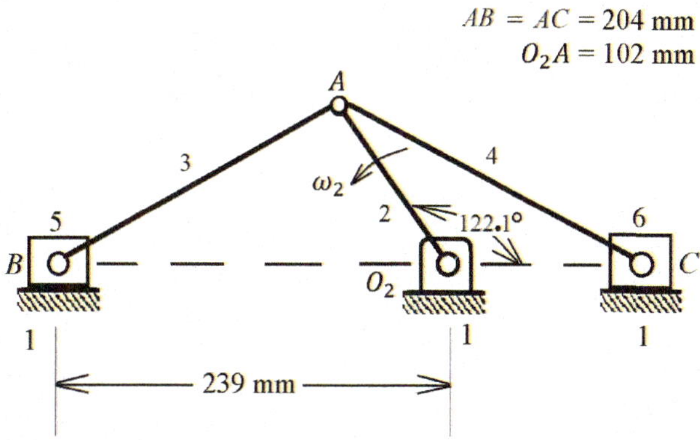

Figure 6.21: Double slider-crank mechanism.

CHAPTER 7

Analysis and Design of Cams

- A cam [1, 2] is a mechanical component/element of a machine that is applied to transmit motion, by direct contact, to another element (known as the follower), through prescribed program of motion.

- A cam mechanism generally contains three elements. The latter are the: cam, follower or follower system, and frame. A typical example is given in Figure 7.1.

- Cams are commonly used in textile machineries, automotive engine assemblies, printing presses, food processing machines, and many automatic devices.

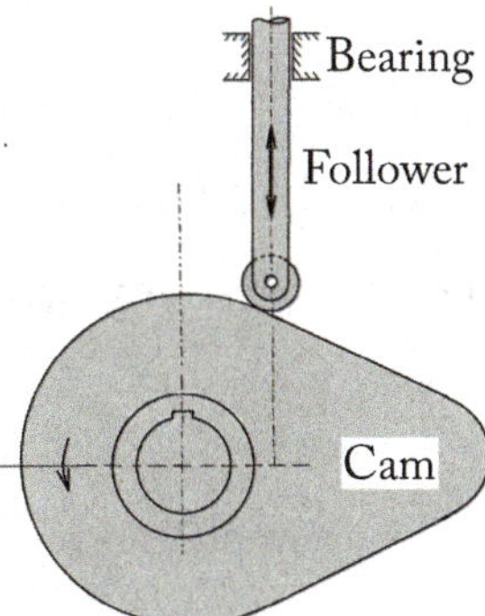

Figure 7.1: A typical cam with follower.

As stated in [3], a cam may be designed in two ways. The first way is to design the cam motion in accordance with the assumed motion of the follower. The second way is to assume the cam shape or profile from which motion characteristics (displacement, velocity, and acceleration) are determined. The first approach is an example of synthesis and it is followed in the present chapter. Before considering the motion design a brief classification of cams is presented in Section 7.1. The classification of followers is introduced in Section 7.2. Follower motion design is included in Section 7.3. Graphical design of cams is introduced in Section 7.4 while analytical design of disk cam with flat-faced follower and disk cam with oscillating roller follower are dealt with in Section 7.5.

Before presenting Section 7.1 the question of *why cams are used in place of common linkages*, such as those introduced in Chapters 3 and 4, may be addressed by way of studying their characteristics in Table 7.1.

Table 7.1: Characteristics of cams and common linkages

Cams	Common Linkages
Can be designed to co-ordinate many motion specifications	Can satisfy relatively limited number of motion requirements
Can be made small and compact	Occupy relatively more space
Dynamic balance can be easily achieved	Dynamic balance involves difficult and complicated analysis
Relatively expensive to made	Less expensive to produce
Dynamic responses are sensitive to manufacturing accuracy of cam contour	Output dynamic responses are less sensitive to slight manufacturing inaccuracy
Subject to surface wear	Joint wear is not critical
Noise level in operation can be relatively high	Produce lower noise level in operation

7.1 CLASSIFICATION OF CAMS

Many types of cams are available in the industries. In general, they may be classified into three categories.

- *Plate or disk cams*

  They are the most common type. An example is given in Figure 7.1.

- *Linear cams*

  Every one of these is characterized by a groove cut into the translating block. The distance of the groove varies from the plane of translation. See, for example, Figure 7.2a.

- *Cylindrical or drum cams*

  In this type of cam a follower rides in the groove which is cut into the cylinder that rotates about its axis. An example is given in Figure 7.2b.

7.2 CLASSIFICATION OF FOLLOWERS

The first consideration in cam design is motion. Next to be considered in cam design are the types and magnitudes of loads. These manifest into the classification of followers since the cam

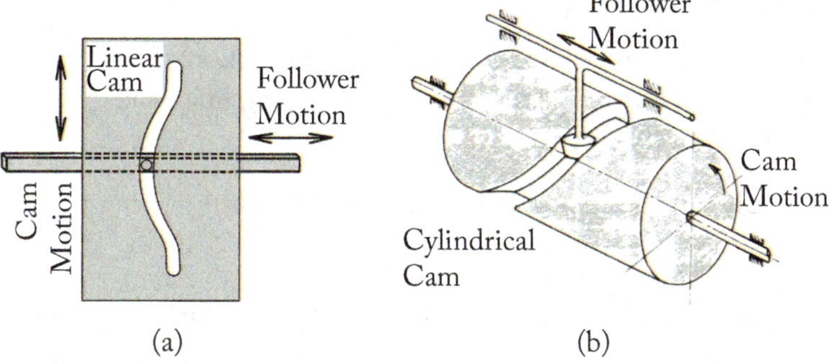

Figure 7.2: Cam mechanisms: (a) linear cam and (b) cylindrical cam.

and follower are in direct contact such that the cam profile can be derived from the path traced by the follower. The followers may be classified by their motion, position, and shape.

7.2.1   FOLLOWER MOTIONS

In general, follower motions are divided into translational and rotational motions. Consequently, there are the so-called translating followers, and swinging arm or pivoted followers, as shown in Figure 7.3.

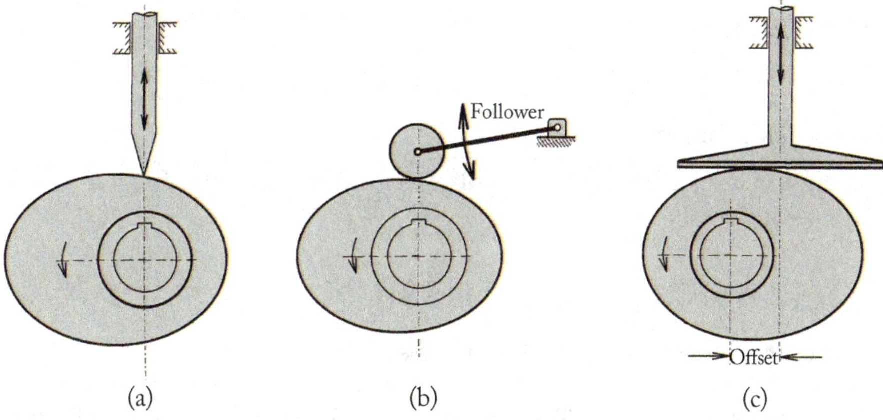

Figure 7.3: Cams with translating and swinging arm followers: (a) in-line knife-edge follower, (b) roller follower, and (c) offset flat-faced follower.

7.2.2   FOLLOWER POSITIONS

In the case of swinging arm or pivoted followers, no distinction of position is necessary since they possess identical kinematics. However, in the translating followers they are divided into the in-line followers and offset followers.

- **In-line followers:** the line of translation extends through the center of rotation of the cam; see Figure 7.3a.

- **Offset followers:** the line of motion is off-set from the center of rotation of the cam; see Figure 7.3c.

7.2.3   FOLLOWER SHAPES

Shapes of followers can be divided into *knife-edge followers*, *roller followers*, *flat-faced followers*, and *spherical-face followers*.

(a) **Knife-edge followers**

Have the simplest shape but wear quickly because of high contact stresses.

(b) **Roller followers**

These are the most frequently applied followers since contact stresses and friction are lower than the knife-edge followers. However, it is possible that they can jam during steep cam displacement motions.

(c) **Flat-faced followers**

This type of followers can be used with steep cam displacement motions.

(d) **Spherical-face followers**

These can also be used with steep cam displacement motions. But friction forces are larger than those of the roller followers.

7.3   FOLLOWER MOTION DESIGN

- Motion of the follower is specified for the task desired.

- Follower displacement diagram is required. A plot of the follower displacement against time or angular displacement is referred to as the *follower displacement diagram*.

Example 7.1

A cam to be designed for a platform that will lift cartons of bottles repeatedly from the lower conveyor belt to an upper conveyor belt is considered. This system is shown in Figure 7.4. The objective of this question is to plot a displacement diagram and determine the required angular velocity of the cam when the motion sequences/schemes of the follower are given as follows.

Segment 1:   Lift 3 cm in 1.3 s,

Segment 2:   Dwell for 0.4 s,

Segment 3:   Fall 1 cm in 0.8 s,

Segment 4:   Dwell 0.5 s, and

Segment 5:   Fall 2 cm in 1.0 s.

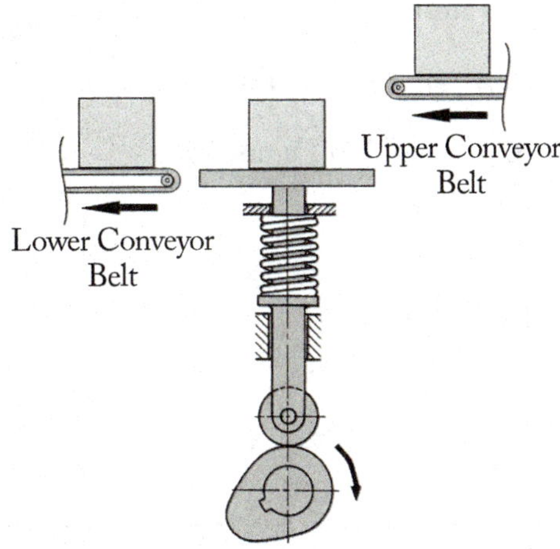

Figure 7.4: Cam in simple automation system.

Solution:
The total time for completing the whole cycle in the above system is

$$t_c = \sum_{i=1}^{5} t_i = (1.3 + 0.4 + 0.8 + 0.5 + 1.0)\ \text{s} = 4.0\ \text{s}.$$

Therefore, the angular velocity of the system is

$$\omega = \frac{1\ \text{rev}}{t_c} = \frac{1}{4.0}\ \frac{\text{rev}}{\text{s}} = 0.250\ \text{rps} = 15.0\ \text{rpm}.$$

Now, therefore, one can obtain the angular displacement for every segment of the motion sequence as presented in the following.
The angular displacement in the first segment (of the follower motion) is

$$\beta_1 = \omega(t_1) = 0.250(1.30)\ \text{rev} = 0.325\ \text{rev} = 117.0°.$$

The angular displacement in the second segment is

$$\beta_2 = \omega(t_2) = 0.250(0.4)\ \text{rev} = 0.10\ \text{rev} = 36.0°.$$

Similarly, the angular displacement in the third segment is

$$\beta_3 = \omega(t_3) = 0.250(0.80)\ \text{rev} = 0.20\ \text{rev} = 72.0°.$$

The angular displacement in the fourth segment is

$$\beta_4 = \omega(t_4) = 0.250(0.50)\ \text{rev} = 0.125\ \text{rev} = 45.0°.$$

Finally, the angular displacement in the fifth segment is

$$\beta_5 = \omega(t_5) = 0.250(1.0)\ \text{rev} = 0.250\ \text{rev} = 90.0°.$$

The results in the foregoing are presented in Figure 7.5.

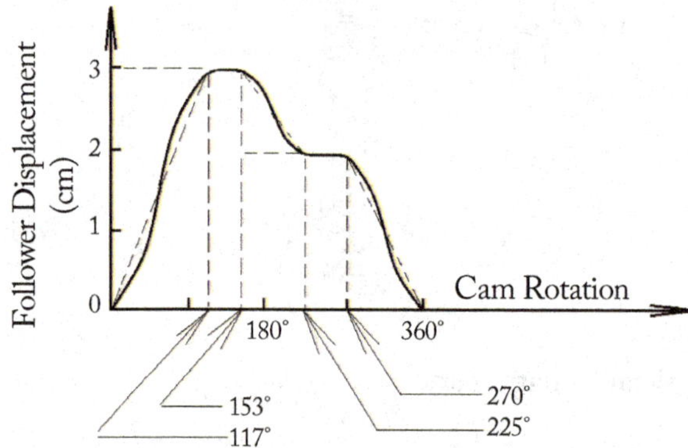

Figure 7.5: Constructed follower displacement.

Recall from the introduction to this chapter that one way to design the cam motion is to assume the motion of the follower. This motion is governed by the sequences/schemes of motions desired.

- The required condition or objective for the motion schemes is to provide smooth accelerations so that jumps may be avoided.

There are many possible motion schemes in the cam design. The simplest ones are the *constant velocity* schemes. These and the cycloidal as well as harmonic motion schemes [4] are included in the following sections.

- A comment on exclusion of procedures for graphical construction of individual cycloidal and harmonic motion schemes is in order. Frequently, these procedures are included in books that deal with similar topics. However, given the computer power and easily available software, such as MatLab or Excel in Microsoft Office suite, computer-generated cycloidal, harmonic, and other motion schemes for follower motion design can be conveniently plotted. Thus, procedures for individual motion scheme graphical construction are not included in the present book.

7.3.1 CONSTANT VELOCITY MOTION SCHEME

This is the simplest follower motion. Its characteristics are given in Table 7.2 and graphical representation is included in Figure 7.6. For consciseness, all follower displacement, velocity, and acceleration all are plotted in one figure. Note that they are of different units.

Table 7.2: Characteristics of constant velocity follower scheme

	Lift	Return
Displacement, s	$L\dfrac{t}{T}=\dfrac{L\theta}{\beta}$	$L\left(1-\dfrac{t}{T}\right)=\dfrac{L\,(1-\theta)}{\beta}$
Velocity, v	$\dfrac{L}{T}=\dfrac{L\omega}{\beta}$	$-\dfrac{L}{T}=-\dfrac{L\omega}{\beta}$
Acceleration, a	0	0

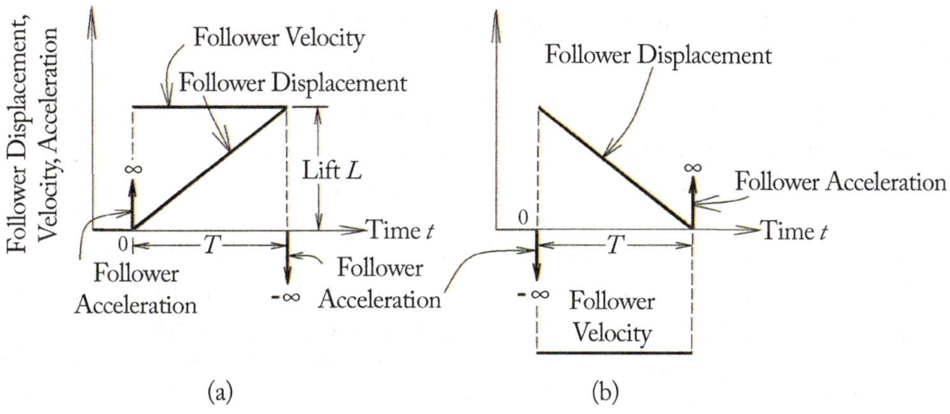

Figure 7.6: Constant velocity motion characteristics: (a) lift and (b) return.

7.3.2 CYCLOIDAL MOTION SCHEMES

Cycloidal motion scheme characteristics are included in Appendix 7A in which the lift and return schemes from [3] are presented. In the latter reference the constant angular velocity ω has been disregarded. The cycloidal motion schemes presented in [3] were adopted from [4].

7.3.3 HARMONIC MOTION SCHEMES

Harmonic motion scheme characteristics are presented in Appendix 7A. Similar to the cycloidal motion schemes the constant angular velocity ω for harmonic motion schemes in [3] has been disregarded. The harmonic motion schemes in [3] were adopted from [4].

Example 7.2
A follower is to be driven from a dwell position through a total lift (rise) L of 100 mm and return (fall) completely to a second dwell. The rise and fall motions take place during $\pi/3$ and $\pi/2$ of cam rotation, respectively. The rotational speed is uniform. The types of motion in these rises and falls are specified as follows:

Segment 1 of cam motion:	Cycloid C-1 lift	$\beta_1 = \pi/6,$
Segment 2 of cam motion:	Harmonic H-2 lift	$\beta_2 = \pi/6,$
Segment 3 of cam motion:	Harmonic H-3 return,	
Segment 4 of cam motion:	Cycloid C-4 return.	

Establish all conditions of displacement, velocity, and acceleration to ensure smooth motion and obtain actual lift equations for each interval of motion.

Solution:
Total lift or rise $L = 100$ mm.
Lift motion: $\beta_1 + \beta_2 = \pi/3$;   return or fall motion: $\beta_3 + \beta_4 = \pi/2$.
Given: $\beta_1 = \beta_2 = \pi/6$.
The follower motion segments according to the given parameters are sketched in Figure 7.7. Note that C-1, H-2, H-3, and C-4 cam motions can be obtained from [3] or [4]. For direct application these motion schemes have been included in Appendix 7A of the present chapter. In this solution they will be listed in details at the appropriate stages for completeness and clear explanation. The analysis starts from point B.

Matching Velocities at B (accelerations both zero)
With reference to Appendix 7A, C-1 has the following equations:

$$s = L\left[\frac{\theta}{\beta} - \frac{1}{\pi}\sin\left(\pi\frac{\theta}{\beta}\right)\right], \quad v = \frac{L}{\beta}\left[1 - \cos\left(\pi\frac{\theta}{\beta}\right)\right], \quad a = \frac{\pi L}{\beta^2}\left[\sin\left(\pi\frac{\theta}{\beta}\right)\right],$$

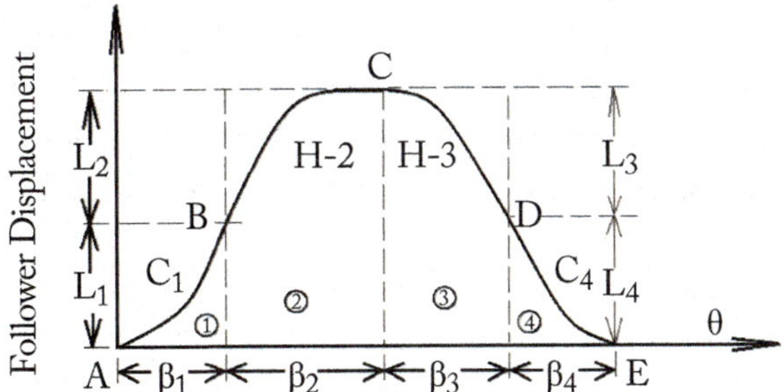

Figure 7.7: Given follower motion.

and H-2 has the following equations:

$$s = L\left[\sin\left(\frac{\pi\theta}{2\beta}\right)\right], \quad v = \frac{\pi L}{2\beta}\left[\cos\left(\frac{\pi\theta}{2\beta}\right)\right], \quad a = -\frac{\pi^2 L}{4\beta^2}\left[\sin\left(\frac{\pi\theta}{2\beta}\right)\right],$$

therefore, for the first segment or C-1 at B: $\theta = \beta_1$ and $\beta = \beta_1$, leading to the result of acceleration, $a = 0$.

Similarly, for the second segment or H-2 at B: $\theta = 0$ and $\beta = \beta_2$, leading to the result of acceleration, $a = 0$. Since accelerations of C-1 and H-2 at B are zero one has to consider the velocities. Applying the velocity equations for C-1 and H-2, one obtains

$$\frac{L1}{\beta_1}(2) = \frac{\pi}{2}\frac{L_2}{\beta_2} \quad \text{or} \quad L_1\beta_2 = \frac{\pi}{4}L_2\beta_1. \tag{7.1}$$

Since

$$L = L_1 + L_2 = L_3 + L_4, \tag{7.2}$$

therefore,

$$L_1\beta_2 = \frac{\pi}{4}(L - L_1)\beta_1 \quad \text{or} \quad L_1\left(\frac{\pi}{4}\beta_1 + \beta_2\right) = \frac{\pi}{4}L\beta_1.$$

Hence,

$$L_1 = L/\left(1 + \frac{4}{\pi}\frac{\beta_2}{\beta_1}\right) = 100/\left(1 + \frac{4}{\pi}\right) \text{ mm} = 43.99 \text{ mm}.$$

Thus, $L_2 = L - L_1 = (100 - 43.99)$ mm $= 56.01$ mm.

Matching accelerations at C (velocities both zero)
From Appendix 7A, H-3 has the following equations:

$$s = L\left[\cos\left(\frac{\pi\theta}{2\beta}\right)\right], \quad v = -\frac{\pi L}{2\beta}\left[\sin\left(\frac{\pi\theta}{2\beta}\right)\right], \quad a = -\frac{\pi^2 L}{4\beta^2}\left[\cos\left(\frac{\pi\theta}{2\beta}\right)\right],$$

therefore, for the second segment or H-2 at $C : \theta = \beta_2$ and $\beta = \beta_2$, leading to the result of velocity, $v = 0$.

Similarly, for the third segment or H-3 at $C : \theta = 0$ and $\beta = \beta_3$, leading to the result for velocity, $v = 0$. Since velocities for H-2 and H-3 at C are zero one has to consider the accelerations. Applying the acceleration equations for H-2 and H-3, one can obtain

$$-\frac{\pi^2 L_2}{4\beta_2^2}(1) = -\frac{\pi^2 L_3}{4\beta_3^2}(1), \quad \text{or} \quad L_2\beta_3^2 = L_3\beta_2^2. \tag{7.3}$$

Matching velocities at D (accelerations both zero)

From Appendix 7A, C-4 has the following equations:

$$s = L\left[1 - \frac{\theta}{\beta} - \frac{1}{\pi}\sin\left(\frac{\pi\theta}{\beta}\right)\right], \quad v = -\frac{L}{\beta}\left[1 + \cos\left(\frac{\pi\theta}{\beta}\right)\right], \quad a = \frac{\pi L}{\beta^2}\left[\sin\left(\frac{\pi\theta}{\beta}\right)\right],$$

therefore, for the third segment or H-3 at $D : \theta = \beta_3$ and $\beta = \beta_3$, leading to the result for acceleration, $a = 0$.

Similarly, for the fourth segment or C-4 at $D : \theta = 0$ and $\beta = \beta_4$, leading to the result for acceleration, $a = 0$. Applying the velocity equations for H-3 and C-4, one has

$$-\frac{\pi L_3}{2\beta_3}(1) = -\frac{L_4}{\beta_4}(2) \quad \text{or} \quad L_4\beta_3 = \frac{\pi}{4}L_3\beta_4. \tag{7.4}$$

Recall,

$$L = L_3 + L_4, \tag{7.5}$$

$$\beta_3 + \beta_4 = \frac{\pi}{2}. \tag{7.6}$$

Solving Equations (7.4), (7.5), and (7.6), one obtains

$$\frac{\pi}{4}\left(\frac{\pi}{2}\right)L_3 = \sqrt{\frac{L_3}{L_2}}\beta_2\left[L + L_3\left(\frac{\pi}{4} - 1\right)\right]$$

or by substituting the given data, one has

$$\frac{\pi}{4}\left(\frac{\pi}{2}\right)L_3 = \sqrt{L_3}(0.06996)(100 - 0.2146L_3).$$

By squaring both sides of the above equation and re-arranging terms, it results

$$L_3^2 - 7483.78L_3 + 217136.6 = 0.$$

Solving this quadratic equation, one obtains

$$L_3 = 28.36 \text{ mm}.$$

Then, from Equation (7.5),

$$L_4 = 100 - L_3 = 71.64 \text{ mm}.$$

From Equation (7.3),

$$\beta_3^2 = \left(\frac{L_3}{L_2}\right)\beta_2^2 = \frac{28.36}{56.01}\left(\frac{\pi}{6}\right)^2.$$

Therefore, $\beta_3 = 0.3726$ rad   or $21.35°$.

From Equation (7.4),

$$\beta_4 = \frac{4}{\pi}\left(\frac{L_4}{L_3}\right)\beta_3 = \frac{4}{\pi}\left(\frac{71.64}{28.36}\right)0.3726 = 1.198 \text{ rad}   \text{or}   68.65°.$$

In fact, since β_3 has been found above and therefore one can apply directly the given data that $\beta_3 + \beta_4 = \pi/2$ to arrive at the same result for β_4. Collecting all the results,

$$L_1 = 43.99 \text{ mm}, \quad \beta_1 = \tfrac{\pi}{6}, \text{ given.}$$
$$L_2 = 56.01 \text{ mm}, \quad \beta_2 = \tfrac{\pi}{6}, \text{ given.}$$
$$L_3 = 28.36 \text{ mm}, \quad \beta_3 = 0.3726 \text{ rad.}$$
$$L_4 = 71.64 \text{ mm}, \quad \beta_4 = 1.198 \text{ rad.}$$

With these parameters the required conditions are satisfied.

7.3.4   REMARKS

The selection of motion schemes to satisfy particular engineering specifications of follower design is made under the following conditions.

- The cycloid gives zero acceleration at both ends of the action. Hence, it can be connected to a dwell at each end. Since the pressure angle is relatively high and the acceleration falls to zero unnecessarily, two cycloids should not be connected together.

- The harmonic motion scheme gives the smallest peak acceleration and pressure angle. Thus, it is recommended when acceleration at both beginning and finishing can be matched to the end of acceleration of the adjacent profiles. Owing to the fact that acceleration at the midpoint is zero, the half-harmonic motion scheme can frequently be applied where a constant-velocity lift follows by an acceleration. The half-harmonic scheme can also be connected to a half-cycloid or to a half-polynomial.

- The eighth-power polynomial motion scheme (not included in this chapter for brevity) has a non-symmetrical acceleration curve and it gives a peak acceleration and pressure angle intermediate between the cycloidal and harmonic motion schemes.

7.4   GRAPHICAL DESIGN OF CAMS

As commented in Section 7.3, procedures for graphical construction of individual cycloidal and harmonic motion schemes were excluded in this book due to the ease of generating them by computer software. This comment is also true for cam profiles that can be designed analytically. However, in some situation where the whole cam profile is desired graphically or for rapid verification of the computer-generated results the following steps for the construction may be helpful.

The following three sections include graphical constructions of three commonly applied cams. They are the disk cam with radial flat-faced follower, disk cam with radial roller follower, and cam with offset roller follower. It is assumed that the follower displacement diagram (FDD) and the radius of the base circle are given in every case. The particular FDD is first divided into equal segments.

7.4.1   DISK CAM WITH RADIAL FLAT-FACED FOLLOWER

Once the FDD has been divided into equal segments the procedure for graphical construction consists of the following steps.

Step 1: Draw the base circle.

Step 2: Draw radial lines from the center of the cam. The numbers of the radial lines, for direct identification purpose, are kept as those in the FDD. For example, with reference to Figure 7.8, the radial line numbered with integer 1 (representing the axis of the shaft of the follower at 30° from the vertical which is sometimes referred to as the home position) corresponds to the same number along the angular displacement or horizontal axis of the FDD. Note that the angular displacement axis intersects with the axis of the follower at its home position at right angle.

Step 3: Measure the follower displacement (FD) for every segment. For instance, in the FDD for this case, the FD segments are s_1, s_2, s_3, and so on. Note that for the present case, the FD at segments 4, 5, and 6 are equal. In other words, $s_4 = s_5 = s_6$.

Step 4: Add s_1, s_2, s_3, and so on to all the corresponding radial lines measuring from the base circle. Thus, for example, s_2 is added to the base circle and measuring along horizontal lines, as shown in Figure 7.8.

Step 5: Draw perpendicular lines to the radial lines at distances equal to the base circle radius plus FD. These perpendicular lines represent the face of the follower.

Step 6: Draw a smooth curve tangent to the flat-faced surfaces of the follower. This smooth curve as shown in Figure 7.8 is the contour/shape/profile of the cam. Note that only one half of the cam contour is shown as the other half can be copied from the first half easily.

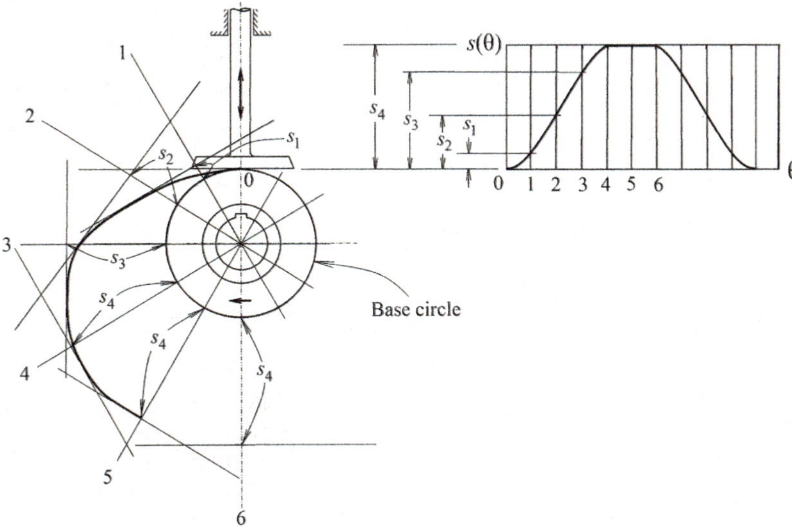

Figure 7.8: Construction of disk cam with radial flat-faced follower.

7.4.2 DISK CAM WITH RADIAL ROLLER FOLLOWER

Once the FDD has been divided into equal segments the procedure for graphical construction consists of the following steps.

Step 1: Draw the base circle and the reference circle with radius $R_r = r_b + r_r$.

Step 2: Draw radial lines from the center of the cam. The numbers of the radial lines are kept as those in the FDD. The angular displacement of horizontal axis of the FDD intersects with the axis of the follower at its home position at right angle.

Step 3: Measure the FD for every segment. Note that for the present case, the FD at segments 4, 5, and 6 are equal such that $s_4 = s_5 = s_6$.

Step 4: Add the FD s_1, s_2, s_3, and so on to their corresponding radial lines at the reference cicle.

Step 5: Draw outlines of rollers, as shown in Figure 7.9.

Step 6: Draw a smooth curve tangent to the outlines of the rollers. This smooth curve, as shown in Figure 7.9, is the contour of the cam. Note that only one half of the cam contour is shown as the other half can be copied from the first half easily.

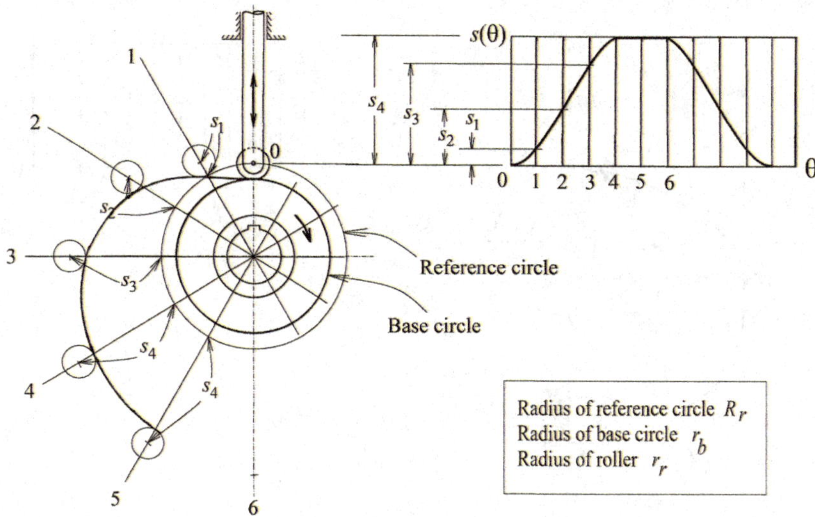

Figure 7.9: Construction of cam with radial roller follower.

7.4.3   CAM WITH OFFSET ROLLER FOLLOWER

Once the FDD has been divided into equal segments the procedure for graphical construction consists of the following steps.

Step 1: Draw the base circle.

Step 2: Draw radial lines from the center of the cam. The numbers of the radial lines are kept as those in the FDD. The angular displacement/horizontal axis of the FDD intersects with the axis of the follower at its home position at right angle.

Step 3: Draw a reference circle whose radius is the sum of the radii of base circle and roller of the follower.

Step 4: Draw an offset circle with radius equal to the offset such that it is tangent to the axis of the follower at its home position.

Step 5: Draw tangents to the offset circle at the radial lines, as shown in Figure 7.10 (the tangents are drawn from the intersection points between the radial lines and offset circles).

Step 6: Measure the FD for every segment and add it along the corresponding tangent to the reference circle. Note again that for the present case, the FD at segments 4, 5, and 6 are equal such that $s_4 = s_5 = s_6$.

Step 7: Draw outlines of rollers, as shown in Figure 7.10.

Step 8: Draw a smooth curve tangent to the outlines of the roller. This smooth curve, as shown in Figure 7.10, is the contour of the cam. Note that only slightly more than one half of the cam contour is shown in Figure 7.10 since the remaining portion of the cam contour can be copied from the plotted portion easily by making use of symmetry.

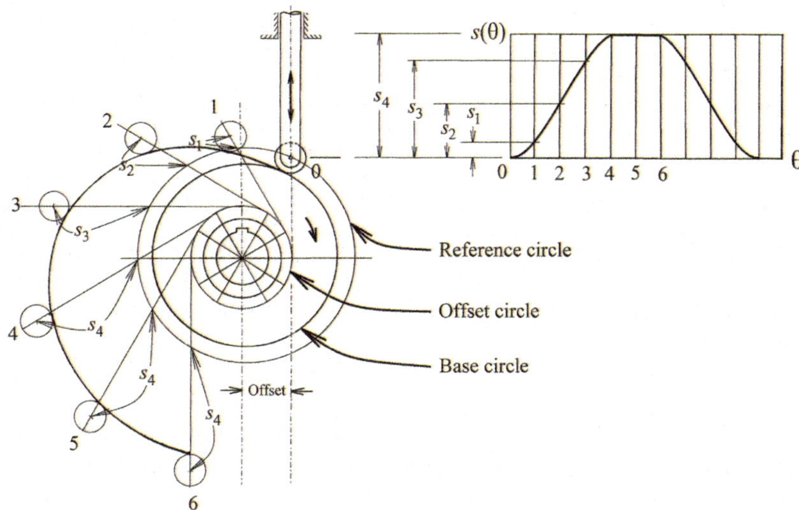

Figure 7.10: Construction of cam with offset roller follower.

7.5 ANALYTICAL APPROACHES TO CAM DESIGN

- For some types of cams with specified cam motions it is possible to design them (the cams) analytically if the rotational speed of the cam is constant.

In what follows, only the representative disk cam with radial flat-faced follower and disk cam with oscillating roller follower are considered.

7.5.1 DISK CAM WITH FLAT-FACED FOLLOWER

A cam with a radial flat-faced follower is shown in Figure 7.11.

- The objective is to find the points of contact between the follower and the cam surface. By connecting these points a cam contour is formed.

The displacement of the follower from the origin of the cam to the center of the follower face is given as

$$R = C + f(\theta), \tag{7.7}$$

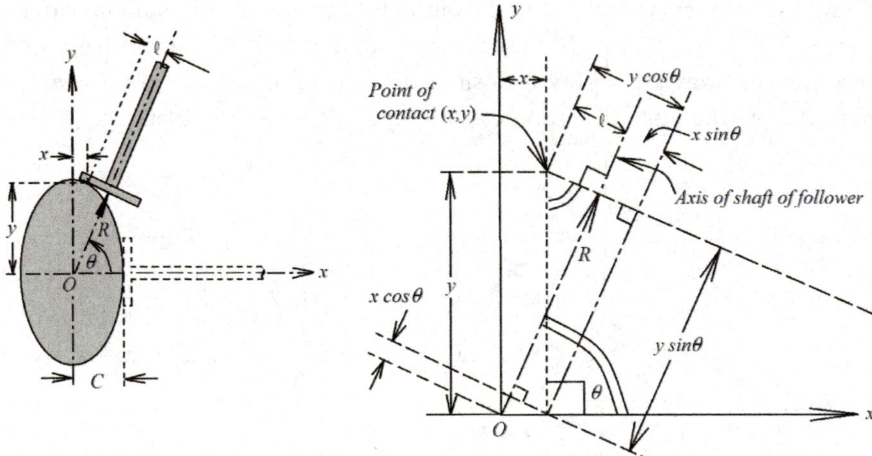

Figure 7.11: Disk cam with flat-faced follower: (a) cam with follower in contact and (b) enlarged view of contacting point and co-ordinates.

where C is the minimum radius of the cam, and $f(\theta)$ the motion of the follower as a function of the angular displacement θ of the cam. Specifically, $f(\theta)$ is s or $s(\theta)$ in Appendix 7A, for example.

With reference to Figure 7.11, one has

$$R = y \sin \theta + x \cos \theta. \tag{7.8}$$

Similarly, the *length of contact* ℓ is given by

$$\ell = y \cos \theta - x \sin \theta. \tag{7.9}$$

From the above two equations, one notes that

$$\ell = \frac{dR}{d\theta} = \frac{d}{d\theta}[C + f(\theta)] = \frac{df(\theta)}{d\theta} = f'(\theta). \tag{7.10}$$

Writing Equations (7.8) and (7.9) in matrix form, one has

$$\begin{pmatrix} R \\ \ell \end{pmatrix} = \begin{bmatrix} \cos \theta & \sin \theta \\ -\sin \theta & \cos \theta \end{bmatrix} \begin{pmatrix} x \\ y \end{pmatrix}.$$

Note that the coefficient matrix is an orthogonal matrix. Thus, its inverse is equal to the transpose of the coefficient matrix. The above matrix equation becomes

$$\begin{pmatrix} x \\ y \end{pmatrix} = \begin{bmatrix} \cos \theta & \sin \theta \\ -\sin \theta & \cos \theta \end{bmatrix}^{-1} \begin{pmatrix} R \\ \ell \end{pmatrix} = \begin{bmatrix} \cos \theta & -\sin \theta \\ \sin \theta & \cos \theta \end{bmatrix} \begin{pmatrix} R \\ \ell \end{pmatrix}.$$

Substituting for R and ℓ, and expanding one has

$$x = [C + f(\theta)]\cos\theta - f'(\theta)\sin\theta, \tag{7.11}$$

$$y = [C + f(\theta)]\sin\theta + f'(\theta)\cos\theta. \tag{7.12}$$

With given θ the corresponding co-ordinates x and y can be obtained so that the disk cam profile can be plotted.

- A cusp or point occurs on the cam surface when both $\frac{dx}{d\theta} = 0$ and $\frac{dy}{d\theta} = 0$. This is because $\frac{dx}{d\theta}$ and $\frac{dy}{d\theta}$ are tangents to the surface of the cam, and when they are equal to 0, the two tangents intersect.

Differentiating Equation (7.11), one has

$$\frac{dx}{d\theta} = [C + f(\theta)](-\sin\theta) + \cos\theta\frac{d[C + f(\theta)]}{d\theta}$$
$$- f''(\theta)\sin\theta - f'(\theta)\frac{d\sin\theta}{d\theta}$$

or

$$\frac{dx}{d\theta} = -[C + f(\theta) + f''(\theta)]\sin\theta, \tag{7.13}$$

where $f''(\theta) = \frac{d^2 f(\theta)}{d\theta^2}$ has been used.

Similarly, one can show that

$$\frac{dy}{d\theta} = -[C + f(\theta) + f''(\theta)]\cos\theta. \tag{7.14}$$

Equations (7.13) and (7.14) can become zero simultaneously only when

$$C + f(\theta) + f''(\theta) = 0.$$

Therefore, to avoid cusps or points of discontinuities the following condition must be satisfied:

$$C + f(\theta) + f''(\theta) > 0. \tag{7.15}$$

- The minimum radius C should be selected in such a way that Equation (7.15) is satisfied. When C is negative it has no practical significance.

- The *facewidth* of the follower face is equal to twice the length of contact $\ell_{\max}$. In practice, the facewidth of $2.10\ell_{\max}$ is frequently used.

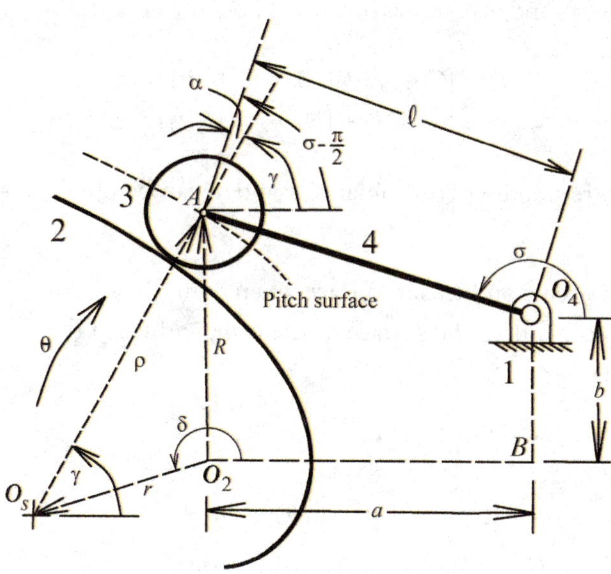

Figure 7.12: Sketch of disk cam and oscillating roller follower.

7.5.2   DISK CAM WITH OSCILLATING ROLLER FOLLOWER

Figure 7.12 shows the sketch of a disk cam with oscillating roller follower. In this figure, the length of the follower shaft ℓ (link 4), r distance between the center O_2 of the cam and center of curvature of the pitch surface O_s, a, and b are all fixed.

Given ℓ, r, a, and b, the cam contour may be generated by the radius of curvature of the pitch surface ρ (subtract radius of roller) in terms of cam angle θ and angular displacement of the follower σ or θ_4 which is

$$\sigma = \sigma_o + f(\theta), \tag{7.16}$$

where $f(\theta)$ is the angular displacement of the follower from the reference angle σ_o (not shown in the figure).

With reference to Figure 7.12, the pressure angle α is given by

$$\alpha = \sigma - \frac{\pi}{2} - \gamma. \tag{7.17}$$

Substituting Equation (7.16) into (7.17),

$$\alpha = [\sigma_o + f(\theta)] - \frac{\pi}{2} - \gamma. \tag{7.18}$$

In order to obtain an equation for angle γ, the approach in [5] based on method of complex variables is adopted in the following. In this approach two independent position equations are written for the center of the roller A.

One equation is written by tracing the path from O_2 to O_s to A (that is, in the clockwise direction) while the other by traveling from O_2 to B to O_4 to A (that is, in the counter-clockwise direction). Thus, for the first path the equation is

$$\vec{R} = re^{i\delta} + \rho e^{i\gamma} = r(\cos\delta + i\sin\delta) + \rho(\cos\gamma + i\sin\gamma). \tag{7.19}$$

The equation for the second route is

$$\vec{R} = a + ib + \ell e^{i\sigma} = a + ib + \ell(\cos\sigma + i\sin\sigma). \tag{7.20}$$

By equating Equation (7.19) to (7.20) and separating the resulting equation into real and imaginary parts, one obtains

$$r\cos\delta + \rho\cos\gamma = a + \ell\cos\sigma, \tag{7.21}$$

$$r\sin\delta + \rho\sin\gamma = b + \ell\sin\sigma. \tag{7.22}$$

For an infinitesimally small rotation of the cam, ρ may be considered to remain constant so that upon differentiating Equations (7.21) and (7.22) with respect to θ,

$$-r\sin\delta\left(\frac{d\delta}{d\theta}\right) - \rho\sin\gamma\left(\frac{d\gamma}{d\theta}\right) = -\ell\sin\sigma\left(\frac{d\sigma}{d\theta}\right), \tag{7.23}$$

$$r\cos\delta\left(\frac{d\delta}{d\theta}\right) + \rho\cos\gamma\left(\frac{d\gamma}{d\theta}\right) = \ell\cos\sigma\left(\frac{d\sigma}{d\theta}\right). \tag{7.24}$$

Since the distance between O_2 and O_s is considered to be fixed therefore $d\delta$ is equal and opposite to $d\theta$ or $\frac{d\delta}{d\theta} = -1$. In addition, recall from Equation (7.16), $\frac{d\sigma}{d\theta} = \frac{df(\theta)}{d\theta} = f'(\theta)$. Thus, Equations (7.23) and (7.24) become

$$r\sin\delta - \rho\sin\gamma\left(\frac{d\gamma}{d\theta}\right) = -\ell f'(\theta)\sin\sigma, \tag{7.25}$$

$$-r\cos\delta + \rho\cos\gamma\left(\frac{d\gamma}{d\theta}\right) = \ell f'(\theta)\cos\sigma. \tag{7.26}$$

Upon eliminating $\frac{d\gamma}{d\theta}$ from Equations (7.25) and (7.26), it results

$$\tan\gamma = \frac{r\sin\delta + \ell f'(\theta)\sin\sigma}{r\cos\delta + \ell f'(\theta)\cos\sigma}.$$

Replacing $r\sin\delta$ from Equation (7.25) and $r\cos\delta$ from (7.26), one has

$$\tan\gamma = \frac{b + \ell\sin\sigma[1 + f'(\theta)]}{a + \ell\cos\sigma[1 + f'(\theta)]} = \frac{D}{C}. \tag{7.27}$$

By making use of this equation, the radius of curvature of the pitch surface is

$$\rho = \left(C^2 + D^2\right)^{3/2} / M \tag{7.28}$$

in which

$$M = \left(C^2 + D^2\right)\left[1 + f'(\theta)\right]$$
$$- (aC + bD)f'(\theta) + (a\sin\sigma - b\cos\sigma)\ell f''(\theta),$$

$$f''(\theta) = \frac{d^2 f(\theta)}{d\theta^2}.$$

It may be appropriate to note that in Equation (3.28) of [3] the second term of M was given as $(aC + bC)f'(\theta)$ and hence Equation (3.28) there is incorrect. The proof of Equation (7.28) is included in Appendix 7B.

- In order to avoid undercutting, the radius of curvature of pitch surface ρ must be larger than the radius of the roller.

7.6   EXERCISES

7.1.   A follower dwells, rises with acceleration, rises with uniform velocity, rises with deceleration, and then dwells again as shown in Figure 7.13. The motion specifications are summarised in Table 7.3.

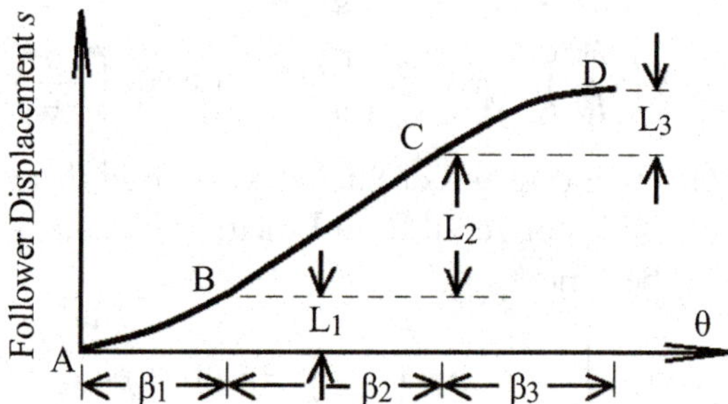

Figure 7.13: Follower displacement diagram.

Table 7.3: Motion specifications for given problem

Point A	Point B	Point C	Point D
$s = 0$	$s = L_1$	$s = L_1 + L_2$	$s = L_1 + L_2 + L_3$
$v = 0$	$v = v_1$	$v = v_1$	$v = 0$
$a = 0$	$a = 0$	$a = 0$	$a = 0$

Choose C-1 for curve AB and C-2 for curve CD for the follower displacement diagram. Determine the relations between L_1, L_2, and L_3, and β_1, β_2, and β_3 to match velocities at points B and C.

7.2. Establish (a) the relation between cam angles β_1, β_2, and lifts L_1, L_2 to match a cycloidal curve C-1 with a harmonic curve H-2, and (b) the relation to match curve H-3 with curve C-4.

7.3. A disk cam turning clockwise at uniform velocity drives a flat-faced radial follower through a rise and fall according to the follower displacement $s = f(\theta) = \frac{L}{2}\left[1 - \cos\left(\frac{2\pi\theta}{\beta}\right)\right]$. The total rise or lift L and the base radius C are both 50 mm. The follower motion takes place over a cam angle of $\beta = \frac{3\pi}{2}$ and dwells for the remainder of the cycle. Construct a graphical layout to establish the surface profile for the cam. Determine graphically the minimum follower face width required.

7.4. Calculate the (x, y) co-ordinates for contact points between the cam and follower given in Exercise 7.3 above. (a) Applying these calculated values plot the layout drawing of the cam, and (b) calculate the minimum follower face width.

Check these results with their corresponding values found graphically in Exercise 7.3.

7.5. A disk cam rotating clockwise at uniform velocity drives a flat-faced radial follower through a lift and return motion according to the following displacement formula

$$s = f(\theta) = \frac{L}{2}\left[1 - \cos\left(\frac{2\pi\theta}{\beta}\right)\right].$$

The total lift L and base radius C are both 50 mm. The follower motion takes place over a cam angle of $\beta = 3\pi/4$.

With this harmonic motion show that for a given value of C and L the cam angle must be limited by the relationship

$$\beta^2 \geq 2\pi^2\left(\frac{L}{C+L}\right),$$

in order to avoid pointing or cusp of the cam surface.

For $C = L = 50$ mm, calculate enough points on the cam surface and plot them to show that a cusp is generated.

7.6. An offset roller follower has the displacement

$$s = f(\theta) = \frac{L}{2}\left[1 - \cos\left(\frac{2\pi\theta}{\beta}\right)\right].$$

The total lift L and base radius C are both 50 mm. The offset is 15 mm and the radius of the roller is 10 mm. The follower motion takes place over a cam angle of $\beta = 3\pi/2$. (a) Derive equations for the co-ordinates (x, y) with their origin located at the center of the cam; and (b) apply the equations derived in part (a), calculate enough points and plot the surface profile of this cam.

REFERENCES

[1] Rothbart, H. A., *Cams*, John Wiley & Sons, 1956. 99

[2] Chen, F. Y., *Mechanics and Design of Cam Mechanisms*, Pergamon Press, 1982. 99

[3] Mabie, H. H. and Ocvirk, F. W., *Mechanisms and Dynamics of Machinery*, 3rd edition-SI Version, John Wiley & Sons, 1978. 99, 106, 118

[4] Kloomok, M. and Muffley, R. V., Plate cam design with emphasis on dynamic effects, *Production Engineering*, February 1955. 104, 106, 121, 123, 125, 127, 129, 131

[5] Raven, F. H., Analytical design of disk cams and three-dimension cams by independent position equations, *American Society of Mechanical Engineers Transactions, Series E*, 26(1), pp. 18–24, 1959. 116

[6] Kloomok, M. and Muffley, R. V., Plate cam design: Radius of curvature, *Production Engineering*, September 1955.

7A   APPENDIX: CYCLOIDAL AND HARMONIC MOTION SCHEMES

For application, cycloidal and harmonic motion schemes are listed in Tables 7A.1–7A.6 while their corresponding plots are presented in Figures 7A.1–7A.12. It may be noted that in these plots $\beta = 1.0$ has been applied.

Table 7A.1: Characteristics of cycloidal motion schemes

	Lift, C-5 [4]	Return, C-6 [4]
Displacement, s	$L\left[\dfrac{\theta}{\beta} - \dfrac{1}{2\pi}\sin\left(\dfrac{2\pi\theta}{\beta}\right)\right]$	$L\left[1 - \dfrac{\theta}{\beta} + \dfrac{1}{2\pi}\sin\left(\dfrac{2\pi\theta}{\beta}\right)\right]$
Velocity, v	$\dfrac{L\omega}{\beta}\left[1 - \cos\left(\dfrac{2\pi\theta}{\beta}\right)\right]$	$-\dfrac{L\omega}{\beta}\left[1 - \cos\left(\dfrac{2\pi\theta}{\beta}\right)\right]$
Acceleration, a	$\dfrac{2\pi L\omega^2}{\beta^2}\sin\left(\dfrac{2\pi\theta}{\beta}\right)$	$-\dfrac{2\pi L\omega^2}{\beta^2}\sin\left(\dfrac{2\pi\theta}{\beta}\right)$

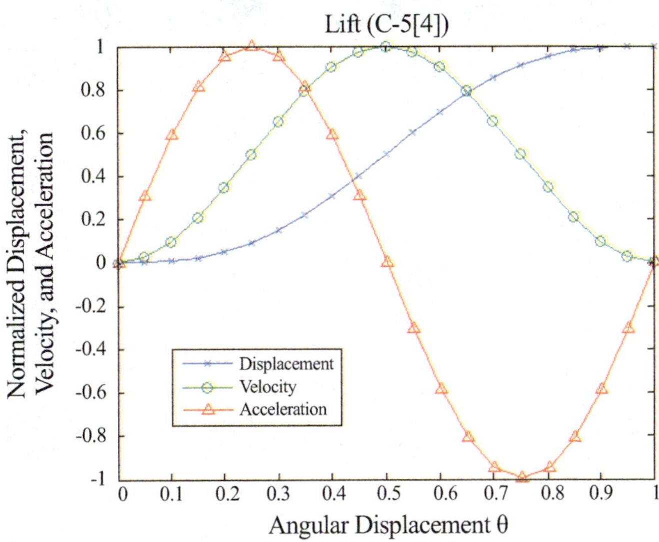

Figure 7A.1: Cycloidal motion schemes: lift, C-5 [4].

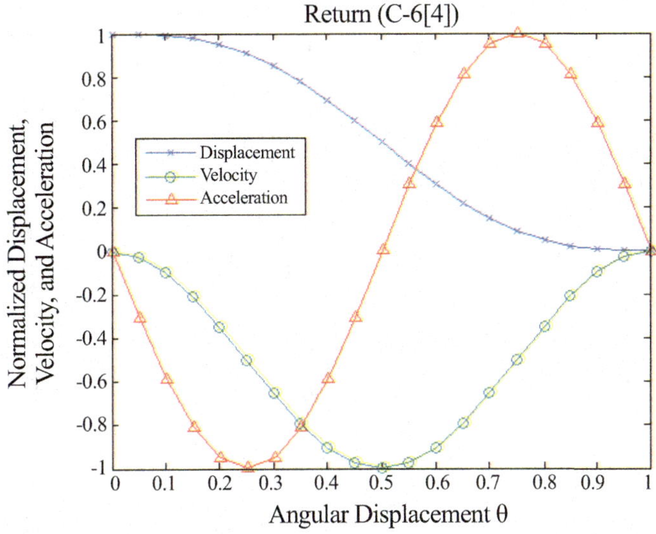

Figure 7A.2: Cycloidal motion schemes: return, C-6 [4].

Table 7A.2: Characteristics of cycloidal motion schemes

	Lift, C-1 [4]	Return, C-4 [4]
Displacement, s	$L\left[\dfrac{\theta}{\beta} - \dfrac{1}{\pi}\sin\left(\dfrac{\pi\theta}{\beta}\right)\right]$	$L\left[1 - \dfrac{\theta}{\beta} - \dfrac{1}{\pi}\sin\left(\dfrac{\pi\theta}{\beta}\right)\right]$
Velocity, v	$\dfrac{L\omega}{\beta}\left[1 - \cos\left(\dfrac{\pi\theta}{\beta}\right)\right]$	$-\dfrac{L\omega}{\beta}\left[1 + \cos\left(\dfrac{\pi\theta}{\beta}\right)\right]$
Acceleration, a	$\dfrac{L\pi\omega^2}{\beta^2}\sin\left(\dfrac{\pi\theta}{\beta}\right)$	$\dfrac{\pi L\omega^2}{\beta^2}\sin\left(\dfrac{\pi\theta}{\beta}\right)$

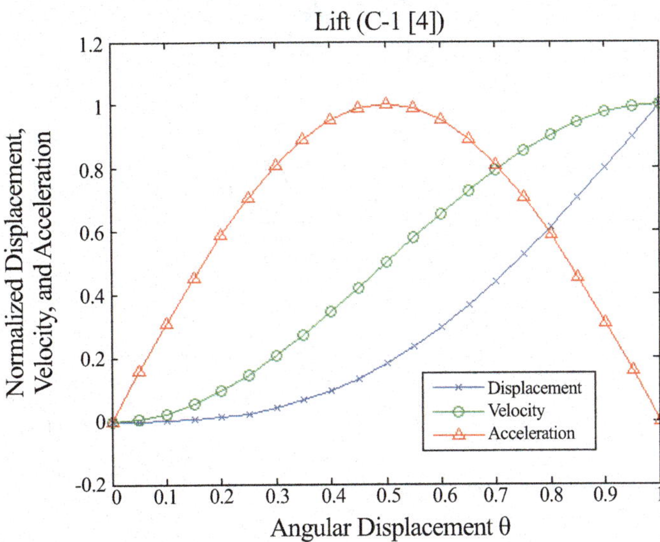

Figure 7A.3: Cycloidal motion schemes: lift, C-1 [4].

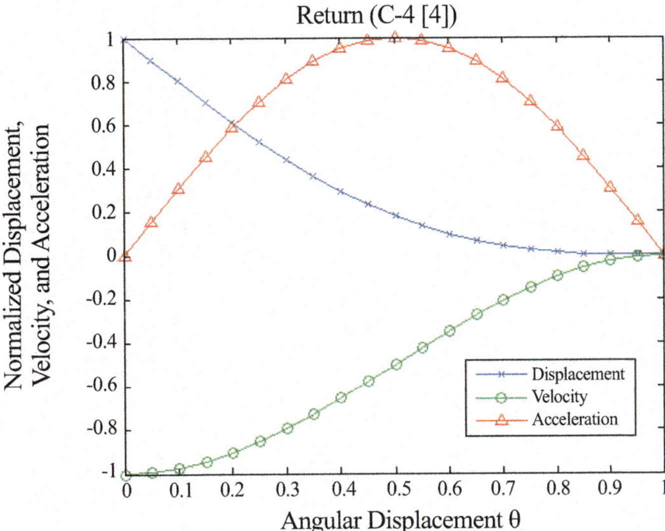

Figure 7A.4: Cycloidal motion schemes: return, C-4 [4].

Table 7A.3: Characteristics of cycloidal motion schemes

	Lift, C-2 [4]	Return, C-3 [4]
Displacement, s	$L\left[\dfrac{\theta}{\beta} + \dfrac{1}{\pi}\sin\left(\dfrac{\pi\theta}{\beta}\right)\right]$	$L\left[1 - \dfrac{\theta}{\beta} + \dfrac{1}{\pi}\sin\left(\dfrac{\pi\theta}{\beta}\right)\right]$
Velocity, v	$\dfrac{L\omega}{\beta}\left[1 + \cos\left(\dfrac{\pi\theta}{\beta}\right)\right]$	$-\dfrac{L\omega}{\beta}\left[1 - \cos\left(\dfrac{\pi\theta}{\beta}\right)\right]$
Acceleration, a	$-\dfrac{\pi L\omega^2}{\beta^2}\sin\left(\dfrac{\pi\theta}{\beta}\right)\right]$	$-\dfrac{\pi L\omega^2}{\beta^2}\sin\left(\dfrac{\pi\theta}{\beta}\right)\right]$

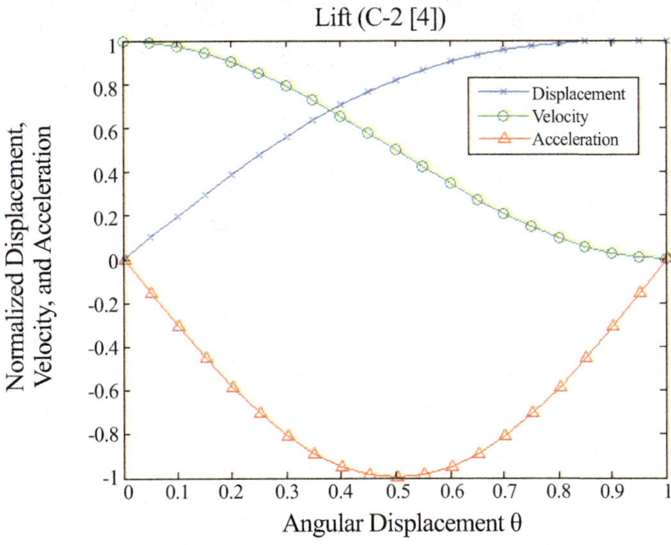

Figure 7A.5: Cycloidal motion schemes: lift, C-2 [4].

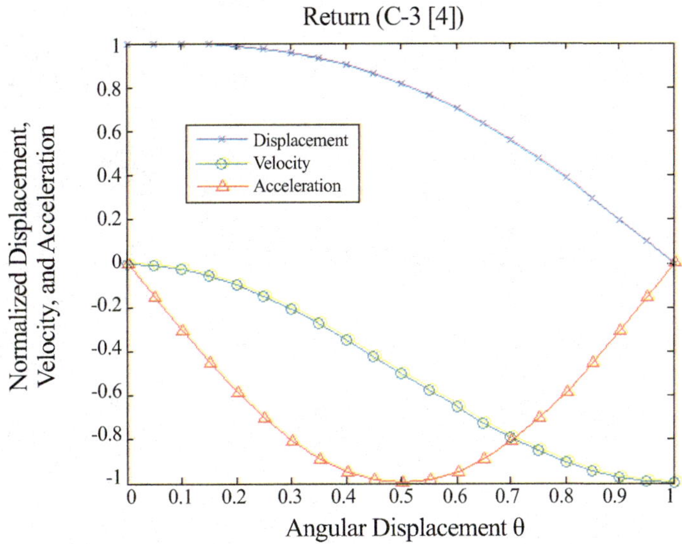

Figure 7A.6: Cycloidal motion schemes: return, C-3 [4].

Table 7A.4: Characteristics of harmonic motion schemes

	Lift, H-5 [4]	Return, H-6 [4]
Displacement, s	$\dfrac{L}{2}\left[1 - \cos\left(\dfrac{\pi\theta}{\beta}\right)\right]$	$\dfrac{L}{2}\left[1 + \cos\left(\dfrac{\pi\theta}{\beta}\right)\right]$
Velocity, v	$\left(\dfrac{\pi L\omega}{2\beta}\right)\sin\left(\dfrac{\pi\theta}{\beta}\right)\right]$	$-\left(\dfrac{\pi L\omega}{2\beta}\right)\sin\left(\dfrac{\pi\theta}{\beta}\right)\right]$
Acceleration, a	$\left(\dfrac{\pi^2 L\omega^2}{2\beta^2}\right)\cos\left(\dfrac{\pi\theta}{\beta}\right)\right]$	$-\left(\dfrac{\pi^2 L\omega^2}{2\beta^2}\right)\cos\left(\dfrac{\pi\theta}{\beta}\right)\right]$

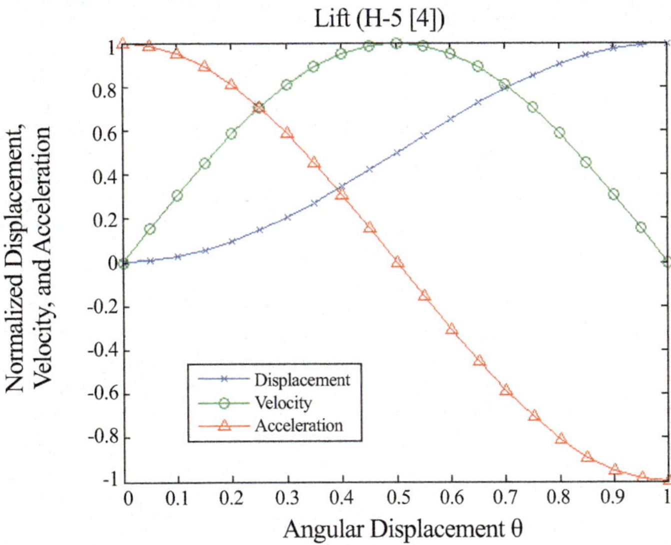

Figure 7A.7: Harmonic motion schemes: lift, H-5 [4].

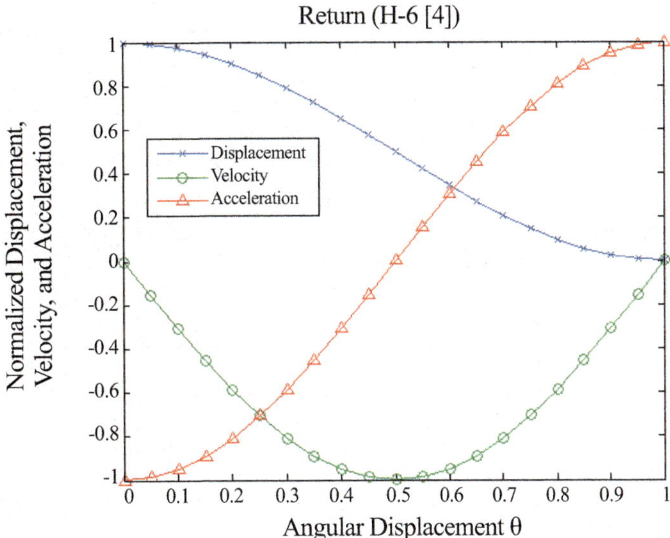

Figure 7A.8: Harmonic motion schemes: return, H-6 [4].

Table 7A.5: Characteristics of harmonic motion schemes

	Lift, H-1 [4]	Return, H-4 [4]
Displacement, s	$L\left[1 - \cos\left(\dfrac{\pi\theta}{2\beta}\right)\right]$	$L\left[1 - \sin\left(\dfrac{\pi\theta}{2\beta}\right)\right]$
Velocity, v	$\left(\dfrac{\pi L\omega}{2\beta}\right)\sin\left(\dfrac{\pi\theta}{2\beta}\right)\Big]$	$-\left(\dfrac{\pi L\omega}{2\beta}\right)\cos\left(\dfrac{\pi\theta}{2\beta}\right)\Big]$
Acceleration, a	$\left(\dfrac{\pi^2 L\omega^2}{4\beta^2}\right)\cos\left(\dfrac{\pi\theta}{2\beta}\right)\Big]$	$\left(\dfrac{\pi^2 L\omega^2}{4\beta^2}\right)\sin\left(\dfrac{\pi\theta}{2\beta}\right)\Big]$

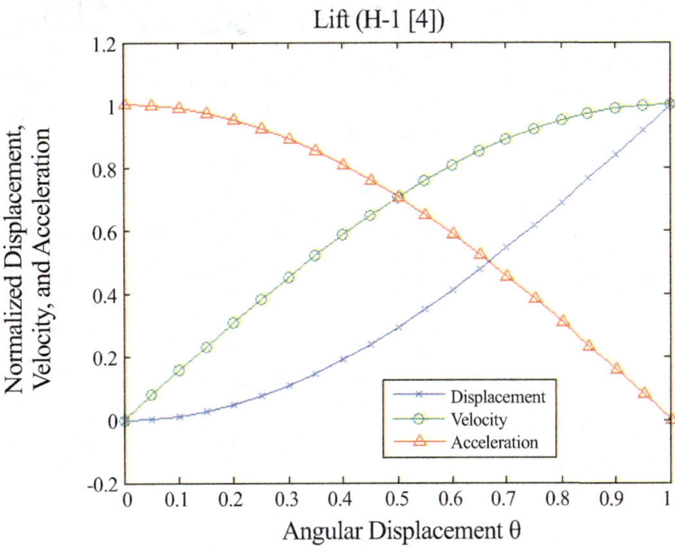

Figure 7A.9: Harmonic motion schemes: lift, H-1 [4].

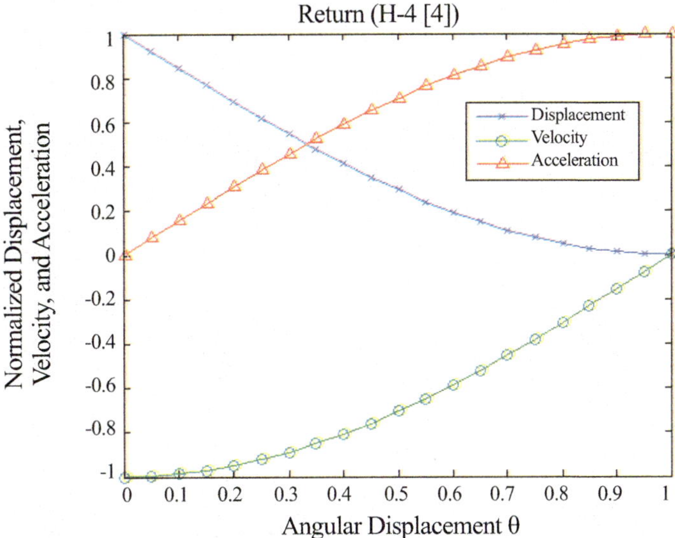

Figure 7A.10: Harmonic motion schemes: return, H-4 [4].

Table 7A.6: Characteristics of harmonic motion schemes

	Lift, H-2 [4]	Return, H-3 [4]
Displacement, s	$L\left[\sin\left(\dfrac{\pi\theta}{2\beta}\right)\right]$	$L\left[\cos\left(\dfrac{\pi\theta}{2\beta}\right)\right]$
Velocity, v	$\left(\dfrac{\pi L\omega}{2\beta}\right)\cos\left(\dfrac{\pi\theta}{2\beta}\right)\Big]$	$-\left(\dfrac{\pi L\omega}{2\beta}\right)\sin\left(\dfrac{\pi\theta}{2\beta}\right)\Big]$
Acceleration, a	$-\left(\dfrac{\pi^2 L\omega^2}{4\beta^2}\right)\sin\left(\dfrac{\pi\theta}{2\beta}\right)\Big]$	$-\left(\dfrac{\pi^2 L\omega^2}{4\beta^2}\right)\cos\left(\dfrac{\pi\theta}{2\beta}\right)\Big]$

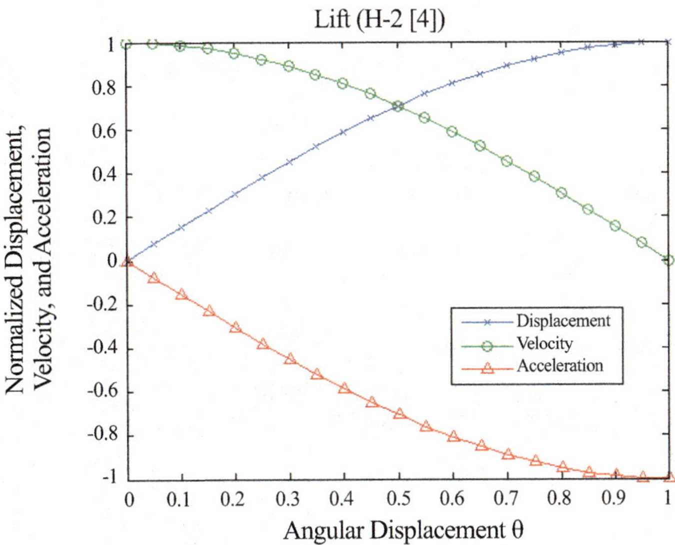

Figure 7A.11: Harmonic motion schemes: lift, H-2 [4].

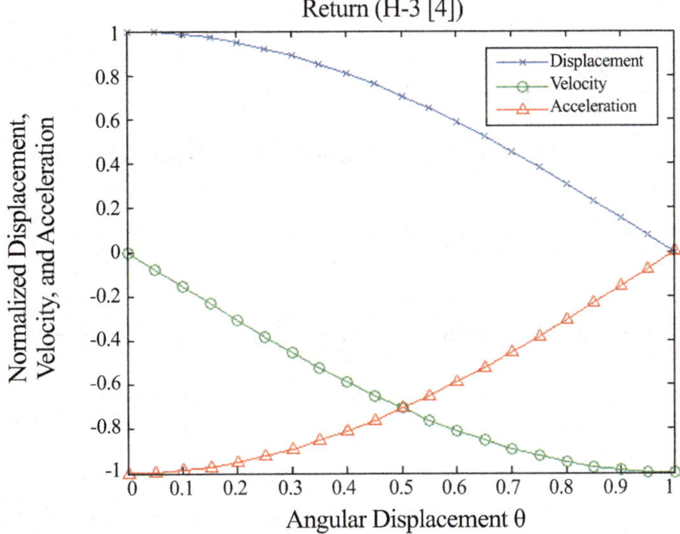

Figure 7A.12: Harmonic motion schemes: return, H-3 [4].

7B   APPENDIX: PROOF OF RADIUS OF CURVATURE OF PITCH SURFACE

From Equation (7.25), one has

$$\rho = \frac{r \sin \delta + \ell f'(\theta) \sin \sigma}{\sin \gamma \left(\frac{d\gamma}{d\theta} \right)}.$$

(7B.1)

Differentiating Equation (7.27) with respect to θ on both sides,

$$\frac{d(\tan \gamma)}{d\theta} = \frac{d(\tan \gamma)}{d\gamma} \left(\frac{d\gamma}{d\theta} \right) = \frac{C \frac{dD}{d\theta} - D \frac{dC}{d\theta}}{C^2}.$$

(7B.2)

Now, consider the first term in the numerator,

$$\frac{dD}{d\theta} = \frac{d\{b + \ell \sin \sigma [1 + f'(\theta)]\}}{d\theta}$$

$$= \ell \left[1 + f'(\theta) \right] \frac{d \sin \sigma}{d\theta} + \ell \sin \sigma \frac{d[1 + f'(\theta)]}{d\theta}$$

$$\frac{dD}{d\theta} = \ell [1 + f'(\theta)] \frac{d \sin \sigma}{d\sigma} \left(\frac{d\sigma}{d\theta} \right) + \ell \sin \sigma f''(\theta)$$

$$= \ell [1 + f'(\theta)] f'(\theta) \cos \sigma + \ell \sin \sigma f''(\theta).$$

Multiplying both sides by C to give

$$C \frac{dD}{d\theta} = C\ell [1 + f'(\theta)] f'(\theta) \cos \sigma + C\ell \sin \sigma f''(\theta).$$

(7B.3)

Similarly, consider the second term in the numerator of Equation (7B.2). By making use of Equation (7.27),

$$\frac{dC}{d\theta} = \frac{d\{a + \ell \cos \sigma [1 + f'(\theta)]\}}{d\theta}$$

$$\frac{dC}{d\theta} = \ell [1 + f'(\theta)] \frac{d \cos \sigma}{d\sigma} \left(\frac{d\sigma}{d\theta} \right) + \ell \cos \sigma \frac{d[1 + f'(\theta)]}{d\theta}$$

$$= -\ell [1 + f'(\theta)] f'(\theta) \sin \sigma + \ell \cos \sigma f''(\theta).$$

Multiplying both sides by D so that

$$D \frac{dC}{d\theta} = -D\ell [1 + f'(\theta)] f'(\theta) \sin \sigma + D\ell \cos \sigma f''(\theta).$$

(7B.4)

Substituting Equations (7B.3) and (7B.4) into (7B.2), and rearranging,

$$\frac{d(\tan \gamma)}{d\theta} = \frac{N_1 + N_2}{C^2},$$

(7B.5)

where $N_1 = \ell f'(\theta)[1 + f'(\theta)](C \cos\sigma + D \sin\sigma)$,   and $N_2 = \ell f''(\theta)(C \sin\sigma - D \cos\sigma)$.
From Equation (7B.2),

$$\frac{d(\tan\gamma)}{d\theta} = \frac{d(\tan\gamma)}{d\gamma}\left(\frac{d\gamma}{d\theta}\right).$$

Therefore,

$$\frac{d\gamma}{d\theta} = \frac{d(\tan\gamma)}{d\theta}\bigg/\frac{d(\tan\gamma)}{d\gamma}.$$

But $\frac{d(\tan\gamma)}{d\gamma} = \sec^2\gamma$. Thus,

$$\frac{d\gamma}{d\theta} = \frac{N_1 + N_2}{C^2 \sec^2\gamma}.$$

Substituting this equation into Equation (7B.1),

$$\rho = \frac{r\sin\delta + \ell f'(\theta)\sin\sigma}{\sin\gamma\left(\frac{N_1+N_2}{C^2\sec^2\gamma}\right)} = \frac{[r\sin\delta + \ell f'(\theta)\sin\sigma]}{(N_1 + N_2)\sin\gamma}C^2\sec^2\gamma. \tag{7B.6}$$

From Equation (7.22),

$$r\sin\delta + \rho\sin\gamma = b + \ell\sin\sigma$$

so that

$$r\sin\delta = b + \ell\sin\sigma - \rho\sin\gamma.$$

Substituting this equation into the numerator term of Equation (7B.6), one has

$$\rho = \frac{[b + \ell\sin\sigma - \rho\sin\gamma + \ell f'(\theta)\sin\sigma]}{(N_1 + N_2)\sin\gamma}C^2\sec^2\gamma$$

$$= \frac{\{b + \ell\sin\sigma[1 + f'(\theta)] - \rho\sin\gamma\}}{(N_1 + N_2)\sin\gamma}C^2\sec^2\gamma.$$

From Equation (7.27), $D = b + \ell\sin\sigma[1 + f'(\theta)]$. Therefore,

$$\rho = \frac{\{D - \rho\sin\gamma\}}{(N_1 + N_2)\sin\gamma}C^2\sec^2\gamma.$$

$$\rho(N_1 + N_2)\sin\gamma = (D - \rho\sin\gamma)C^2\sec^2\gamma.$$

Expanding and rearranging,

$$\rho(N_1 + N_2)\sin\gamma + \rho\sin\gamma\, C^2\sec^2\gamma = DC^2\sec^2\gamma.$$

Therefore,

$$\rho = \frac{DC^2\sec^2\gamma}{(N_1 + N_2)\sin\gamma + \sin\gamma\, C^2\sec^2\gamma}. \tag{7B.7}$$

But $\sin \gamma = \sqrt{1 - \cos^2 \gamma} = \cos \gamma \sqrt{\frac{1}{\cos^2 \gamma} - 1} = \frac{1}{\sec \gamma} \sqrt{\sec^2 \gamma - 1}$.

Since $\sec^2 \gamma - 1 = \tan^2 \gamma$, hence $\sin \gamma = \frac{1}{\sec \gamma} \tan \gamma$. By making use of Equation (7.27),

$$\sin \gamma = \frac{1}{\sec \gamma} \tan \gamma = \frac{1}{\sqrt{\tan^2 \gamma + 1}} \tan \gamma = \frac{1}{\sqrt{\left(\frac{D}{C}\right)^2 + 1}} \left(\frac{D}{C}\right).$$

Therefore,

$$\sin \gamma = \left[\left(\frac{D}{C}\right)^2 + 1\right]^{-1/2} \left(\frac{D}{C}\right) = \frac{D}{\sqrt{C^2 + D^2}}. \tag{7B.8}$$

In addition,

$$\sec^2 \gamma = \tan^2 \gamma + 1 = \left(\frac{D}{C}\right)^2 + 1 = \frac{1}{C^2} \left(C^2 + D^2\right). \tag{7B.9}$$

Substituting Equations (7B.8) and (7B.9) into (7B.7), and simplifying, it gives

$$\rho = \frac{\left(C^2 + D^2\right)^{3/2}}{N_1 + N_2 + C^2 + D^2}, \tag{7B.10}$$

where N_1 and N_2 have been defined in Equation (7B.5).

From the numerator and denominator terms of Equation (7.27), one has

$$\sin \sigma = \frac{D - b}{\ell[1 + f'(\theta)]}, \qquad \cos \sigma = \frac{C - a}{\ell[1 + f'(\theta)]}.$$

Therefore,

$$N_1 = \ell f'(\theta)[1 + f'(\theta)] \left\{ \frac{C(C - a)}{\ell[1 + f'(\theta)]} + \frac{D(D - b)}{\ell[1 + f'(\theta)]} \right\}$$
$$= f'(\theta)\left(C^2 + D^2\right) - f'(\theta)(Ca + Db), \quad \text{and}$$
$$N_2 = \ell f''(\theta)(C \sin \sigma - D \cos \sigma).$$

Recall, from Equation (7.27),

$$C \sin \sigma = \{a + \ell \cos \sigma[1 + f'(\theta)]\} \sin \sigma,$$
$$D \cos \sigma = \{b + \ell \sin \sigma[1 + f'(\theta)]\} \cos \sigma.$$

Therefore,

$$N_2 = \ell f''(\theta)(a \sin \sigma - b \cos \sigma).$$

Substituting N_1 and N_2 into Equation (7B.10), and after simplifying, one obtains

$$\rho = \frac{\left(C^2 + D^2\right)^{3/2}}{M}, \tag{7B.11}$$

where $M = \left(C^2 + D^2\right)\left[1 + f'(\theta)\right]$

$$-(aC + bD)f'(\theta) + (a\sin\sigma - b\cos\sigma)\ell f''(\theta).$$

Equation (7B.11) is Equation (7.28).

C H A P T E R   8

Analysis and Design of Gears

- A gear [1] is a mechanism that can be used to change the speed or direction of motion from one shaft to the other.

- Typical applications of gears are in cars and many other mechanical systems.

- Fundamental law of gearing: *The profile (shape) of the teeth of a gear must be such that the common normal at the point of contact between two teeth always passes through a fixed point (the pitch point) on the line of centers.*

When the above law is satisfied the gears in mesh are said to produce *conjugate action* and will maintain a constant angular velocity ratio.

Various types of gears commonly used are briefly introduced in Section 8.1 while the terminology and spur gears are included in Section 8.2. Gear trains are presented in Section 8.3. Application of gear trains in automotive transmission analysis is introduced in Section 8.4.

8.1   COMMON TYPES OF GEARS IN PRACTICE

In this section gears commonly used in machinery are briefly introduced. These are the spur, helical, racks, internal, bevel, miter, screw, and worm gears.

(a) *Spur gears*

In this type of gears the teeth are straight and parallel to the shaft axes. They transmit power and motion between two rotating parallel shafts.

Characteristics: Easy to manufacture, no axial force, relatively high quality of this type of gears can be produced. It is the most commonly applied type. They are used as transmission components.

(b) *Helical gears*

In this type of gears the teeth are twisted oblique to the gear axis. The hand of helix is designated as either left or right. Right- and left-hand helical gears mate as a set. However, they have the same helix angle.

Characteristics: Having higher strength compared with spur gears, and producing thrust forces in the axial directions. Comparing with spur gears, helical gears have relatively less noise and vibration during operation. They are used as transmission components, in automobiles, and in speed reduction processes.

(c) *Racks*

A rack is a bar containing teeth on one face for meshing with a gear, referred to as pinion or pinion gear. Basically, a rack is a gear of infinite diameter.

A rack changes a rotary motion into a rectilinear motion. They are used as a transfer system for machine tools, printing presses, robots, and so on.

(d) *Internal gears*

An internal gear is an annular gear having teeth on the inner surface of its rim. An internal gear is always meshes with the external gear.

In the meshing of an internal gear with an external gear the rotations in both gears are in the same direction. For a large internal gear in mesh with a small external gear, care has to be taken to address the issue of interferences (see, later in this chapter for definition) which are likely to occur. The machine designed is compact. Internal gears are used in planetary gear drive of high velocity reduction ratio, clutches, and so on.

(e) *Bevel gears*

Every one of a pair of gears used to connect two shafts whose axes intersect, and the pitch surfaces are in the forms of cones. Gear teeth are cut along the pitch cone. Bevel gears are classified, based on the tooth trace, into straight bevel gears and spiral bevel gears.

Characteristics: Providing higher contact ratio and strength, and having better efficiency of transmission with reduced noise. They are used in automobiles, tractors, final reduction gearing for ships, for example.

(f) *Miter gears*

These constitute a special class of bevel gears where the shafts intersect at 90° and produce 1:1 gear ratio.

They have similar characteristics to those of bevel gears. Applications are similar to those of bevel gears with specific conditions stated above.

(g) *Screw gears*

These are helical gears that transmit power from one shaft to another, non-parallel, non-intersecting shaft.

Characteristics: Used in speed reduction and/or a multiplying gear, in sliding contact these gears tend to wear, and not suitable for high horsepower transmission. They are used in automatic machines that provide intricate motion, and as driving gears for automobiles.

(h) *Worm gears*

These are gears that usually applied in pair. A worm is a resulting gear when a tooth on a helical gear makes a complete revolution on the pitch cylinder. The mating gear for a worm is called a worm gear or worm wheel. A worm and worm gear are applied to connect nonparallel, nonintersecting shafts usually at right angles.

Characteristics: Being able to provide large velocity reduction ratios for a given center distance, giving quiet and smooth motion. They are applied in self-locking mechanisms (anti-reversing gear devices), such as indexing devices, chain blocks.

8.2 SPUR GEARS AND TERMINOLOGY

The use of this type of gears is common. Before they are considered in Section 8.2.2, terminology and related equations are first presented in the following section.

8.2.1 TERMINOLOGY

Terminology of gears is included in Figures 8.1 and 8.2. In particular, the *circular pitch p* is the distance measured along the pitch circle from a point on one tooth to the corresponding point of the adjacent tooth of the gear. It is defined as

$$p = \frac{\pi D}{N},$$

(8.1)

where D is the diameter of the pitch circle and N the total number of teeth in the gear. The *pitch circle* of a gear is that traced by the *pitch point* of the gear. The *pitch point* is the point of contact of the two pitch circles from two gears in mesh.

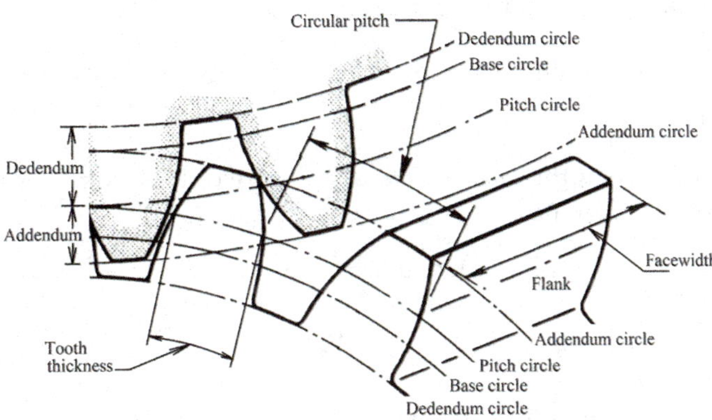

Figure 8.1: Terminology of gears.

- *Diametral pitch P_d* or simply called *pitch* is defined as the ratio of N to D. That is,

$$P_d = \frac{N}{D}.$$

(8.2)

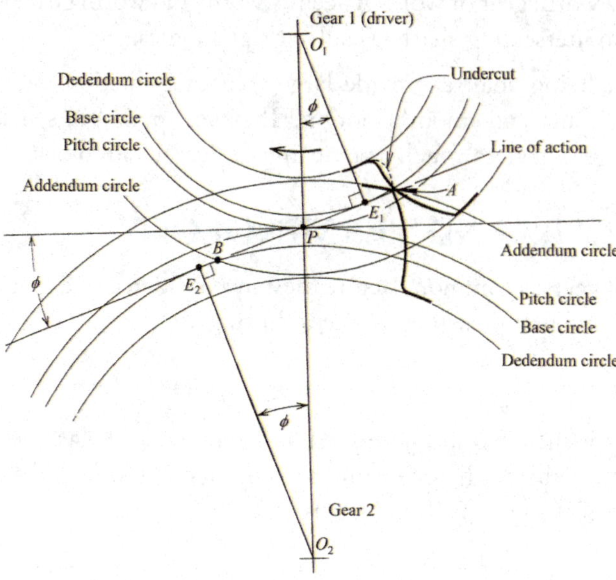

Figure 8.2: Line of action, pressure angle and undercut of gears in mesh.

This is a standard for tooth size specifications. It is expressed in imperial units.

- Note that mating gears must have the same diametral pitch.

- While the diametral pitch is a standard for tooth size specification it can not be measured directly from the gear.

- In the S.I. unit system the *module* $m = D/N$ is used. However, it is not directly equal to the reciprocal of P_d. In fact, $m = 25.4/P_d$.

From Equations (8.1) and (8.2), one has

$$pP_d = \pi. \tag{8.3}$$

Typical diametral pitches are included in Table 8.1.

Most gear tooth parameters are related to the diametral pitch P_d and pressure angle ϕ. Their relationships are provided in Table 8.2.

Example 8.1

A 20° full-depth, involute spur gear with 38 teeth has a diametral pitch of 12. Obtain the diameter of the pitch circle, and base circle.

Table 8.1: Standard diametral pitches

Coarse Pitch		Fine Pitch	
2	6	20	80
2.25	8	24	96
2.5	10	32	120
3	12	40	150
4	16	48	200
		64	

Table 8.2: Standardized relationship of gear tooth parameters

	Coarse Pitch (1-19.99 P_d)* 20° or 25° Full Depth	Fine Pitch (20-200 P_d)@ 20° Full Depth
Addendum, a	$1.000/P_d$	$1.000/P_d$
Dedendum, b	$1.250/P_d$	$1.200/P_d + 0.002$(min)
Clearance, c ($b - a$)	$0.250/P_d$	$0.200/P_d + 0.002$(min)#
Working depth	$2a$	$2a$
Whole depth	$a + b$	$a + b$
Fillet radius of basic rack	$0.300/P_d$	Not given
Tooth thickness	$1.5708/P_d$	$1.5708/P_d$

* American Gear Manufacturers Association (AGMA) 201.02 August 1974.

@ AGMA 207.06 November 1977.

For shaved or ground teeth, $c = 0.350/P_d + 0.002$(min).

Solution:

The pitch diameter $D = N/P_d = 38/12$ in $= 3.17$ in.

The diameter of the base circle is

$$D_b = D \cos \phi = 3.17 \cos 20° = 2.979 \, \text{in.}$$

8.2.2   SPUR GEARS AND DESIGN PARAMETERS

Spur gears are the simplest and most commonly used gears. They are also called involute spur gears because of their tooth shape. The involute tooth shape, properties of which are presented in

Appendix 8A, has several advantages over other possible shapes. The most important advantages are its ease of manufacture, and the fact that the center distance between two involute gears may vary without changing the velocity ratio. Spur gears are used to transmit motion between parallel shafts. Their teeth are parallel to the axis of rotation.

- A rack is a special case of spur gear which can be viewed as a spur gear with infinitely large diameter.

- A translating motion can be produced by a mating rack and spur gear.

The spur gears listed in gear manufacturers' catalogs are *standard gears*. The term *standard gears* is frequently used to mean that the ratio of the number of teeth to pitch diameter is one of the standard values of diametral pitch, and that the tooth thickness must be equal to the tooth space, which equals one-half of the circular pitch [1]. The main important design considerations such as the center distance, contact ratio, and interference are included in the following. Other design considerations such as backlash, and operating pressure angle [1] are not included in this chapter for brevity.

Center Distance

The center distance is defined as the center to center distance between two mating gears. There are two types of center distance. One is associated with external gears and the other is associated with internal gears.

For external gears, the center distance is defined as

$$C_e = R_1 + R_2 = \frac{D_1 + D_2}{2} = \frac{N_1 + N_2}{2P_d}. \tag{8.4}$$

For internal gears, the center distance is defined as

$$C_i = R_2 - R_1 = \frac{D_2 - D_1}{2} = \frac{N_2 - N_1}{2P_d}. \tag{8.5}$$

Example 8.2

Two 6-pitch 20° full-depth gears are used in a chemical reactor mixer. The gears transmit power from a small engine to the mixing drum. The small gear (or pinion) has 18 teeth while the gear (or bull gear) has 32 teeth. Find the center distance.

Solution:

The pitch diameters of both gears are

$$D_1 = \frac{N_1}{P_d} = \frac{18}{6} \text{ in} = 0.3 \text{ in}, \qquad D_2 = \frac{N_2}{P_d} = \frac{32}{6} \text{ in} = 5.33 \text{ in}.$$

Since the gears are external, the center distance becomes

$$C = \frac{D_1 + D_2}{2} = 4.17\,\text{in.}$$

Contact Ratio

In a mating gear system the contact ratio is the average number of teeth that are in contact at any instant.

- A contact ratio of 1.4 means that one pair of teeth is always in contact and a second pair of teeth is in contact 40% of the time.

- Typical contact ratio of 1.4 or 1.5 implies robust design.

  The contact ratio is defined as

$$m_p = \frac{Z}{P_b}, \tag{8.6}$$

  where Z is the *length of action*, and P_b is the base pitch.

- The *line of action* is the common normal to the two involute surfaces and is tangent to the two base circles (see Figure 8.3).

  With reference to Figure 8.3, the *length of action* is expressed as

$$\begin{aligned}
Z = AB &= (AE_2 - PE_2) + (BE_1 - PE_1) \\
&= (AE_2 + BE_1) - (PE_2 + PE_1) \\
&= (AE_2 + BE_1) - (E_2 E_1)
\end{aligned}$$

in which

- A is the point where contact begins,

- B is the point where contact ends,

- E_1, and E_2 are points of tangency of the line of action and base circles (also known as the interference points).

But

$$BE_1 = \sqrt{(R_{o1})^2 - (R_{b1})^2}, \quad AE_2 = \sqrt{(R_{o2})^2 - (R_{b2})^2},$$

and

$$E_2 E_1 = R_2 \sin\phi + R_1 \sin\phi.$$

Therefore, with reference to the geometry in the above figure, one has

$$Z = \sqrt{(R_{o1})^2 - (R_{b1})^2} + \sqrt{(R_{o2})^2 - (R_{b2})^2} + R_2 \sin\phi + R_1 \sin\phi$$

or

$$Z = \sqrt{(R_{o1})^2 - (R_{b1})^2} + \sqrt{(R_{o2})^2 - (R_{b2})^2} + C \sin \phi \qquad (8.7)$$

since the center distance C is equal to the sum of the radii of the pitch circles.

The base pitch is defined by

$$P_b = \frac{2\pi R_b}{N}, \qquad (8.8)$$

where R_b is the radius of the base circle.

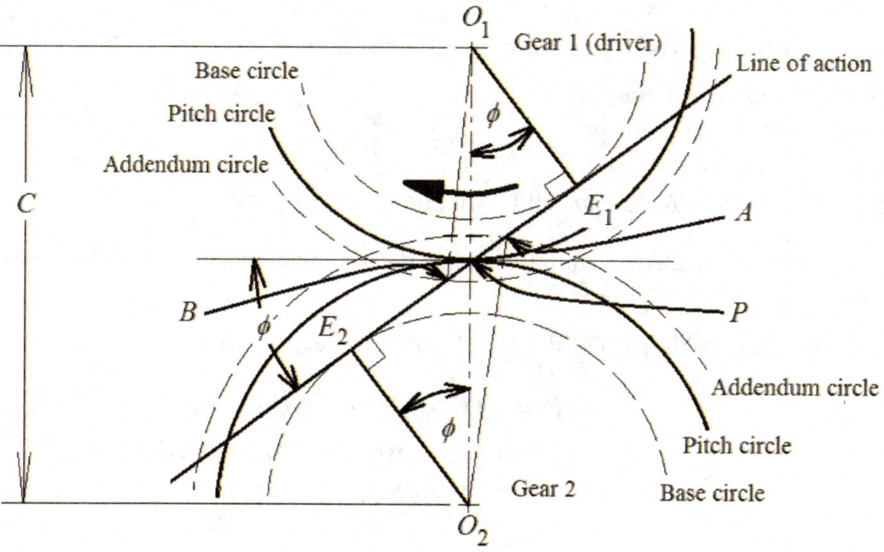

Figure 8.3: Length of action of gears in mesh.

Interference (in Involute Gears)

When a gear with a small number of teeth and small pressure angle is constructed the dedendum circle is considerably smaller than the base circle of the involute such that the tooth between the base circle and the addendum is not an involute. When this portion of the tooth is in contact with that of the mating gear the *fundamental law of gearing* will be violated (because the condition of constant velocity ratio is not satisfied). This condition is known as *interference*.

It occurs frequently when a small gear mates with a much larger one.

- Interference is the most serious drawback in using involute gears. When it occurs the tip of the driven tooth undercuts or gouges out the flank of the driving tooth. This will weaken the strength of the tooth.

- Objective in the following is to derive a relation that provides the condition of no interference.

- The interference condition can be avoided providing that the addendum circle of either gear does not intersect the line of action beyond the interference points.

With reference to Figure 8.3, in order to prevent interference, in the limit (that is, points A and B can not be beyond the interference points E_1 and E_2), points A and B coincide with the interference points E_1 and E_2, respectively. Then, the outside radius of gear 2 can be written as

$$R_{o2} = \sqrt{(R_{b2})^2 + (C \sin \phi)^2} \tag{8.9}$$

since

$$(R_{o2})^2 = (R_{b2})^2 + (E_2 E_1)^2, \quad E_2 E_1 = PE_2 + PE_1,$$
$$PE_1 = R_1 \sin \phi, \qquad PE_2 = R_2 \sin \phi,$$
$$PE_2 + PE_1 = R_2 \sin \phi + R_1 \sin \phi = C \sin \phi.$$

But by definition,

$$R_{o2} = R_2 + a = \frac{N_2 + 2k}{2P_d}, \tag{8.10}$$

$$R_{b2} = R_2 \cos \phi = \frac{N_2}{2P_d} \cos \phi, \tag{8.11}$$

$$C = R_1 + R_2 = \frac{N_1 + N_2}{2P_d}. \tag{8.12}$$

Substituting Equations (8.10), (8.11) and (8.12) into (8.9), one obtains

$$\frac{N_2 + 2k}{2P_d} = \sqrt{\left(\frac{N_2}{2P_d} \cos \phi\right)^2 + \left(\frac{N_1 + N_2}{2P_d}\right)^2 \sin^2 \phi}.$$

Squaring both sides and deleting the common denominator terms, it becomes

$$(N_2 + 2k)^2 = (N_2 \cos \phi)^2 + (N_1 + N_2)^2 \sin^2 \phi.$$

Expanding and simplifying,

$$N_2^2 + 4k^2 + 4kN_2 = N_2^2 \cos^2 \phi + N_1^2 \sin^2 \phi + N_2^2 \sin^2 \phi + 2N_1 N_2 \sin^2 \phi,$$
$$4k^2 + 4kN_2 = N_1^2 \sin^2 \phi + 2N_1 N_2 \sin^2 \phi.$$

Re-arranging,

$$4kN_2 - 2N_1 N_2 \sin^2 \phi = N_1^2 \sin^2 \phi - 4k^2.$$

Therefore,

$$N_2 = \frac{N_1^2 \sin^2 \phi - 4k^2}{4k - 2N_1 \sin^2 \phi}. \tag{8.13}$$

This equation gives the limiting value for a driven gear when no interference is expected.

It means that if one has designed a gear system in such a way that N_2 (selected from the manufacturer's catalog, say) is smaller than the calculated N_2 then no interference will occur.

- In other words, select N_2 which is less than the calculated N_2 then no interference will occur.

8.3   GEAR TRAINS

In many applications it is necessary to have large velocity reduction or change the direction of rotation of a gear without changing its angular velocity. This can be accomplished by using a combination of gears which is called a *gear train*. Formally, a *gear train* is a mechanism or system that consists of several mating gears.

In general, there are gear trains that have their gear centers attached to fixed bodies and those whose centers can be fixed and can be allowed to move. Typical gear trains with fixed centers, and planetary gear trains whose centers may be allowed to move are included in Sections 8.3.1 and 8.3.2, respectively.

8.3.1   GEAR TRAINS WITH FIXED GEAR CENTERS AND TRAIN VALUE

Consider a *compound gear* that is defined as one in which two or more gears are fixed to the same shaft.

- Gear trains are used to achieve large speed change. An example can be found in a garage door opener.

When multiple gears are arranged in a series the overall velocity ratio is known as the *train value* and it is defined as

$$\mu = \frac{\omega_{in}}{\omega_{out}} = \alpha_{12}\alpha_{23}\alpha_{34}\ldots, \tag{8.14}$$

where α_{ij} is the velocity ratio with i denoting the driver gear and j being the driven gear. That is,

$$\alpha_{ij} = \frac{\omega_i}{\omega_j} = \frac{R_j}{R_i},$$

in which R_i is the radius of the pitch circle of gear i, for example.

Example 8.3

Figure 8.4 shows a gear train, in which all gear centers are attached to a fixed reference or fixed frame, with the following properties:

Gear 1: $N_1 = 10$ teeth, and $P_d = 10$,

Gear 2: $D_2 = 2.5$ in,

Gear 3: $N_3 = 15$ teeth,

Gear 4: $D_4 = 3.0$ in, and $P_d = 10$,

Gear 5: $D_5 = 1.5$ in, and $P_d = 10$, and

Gear 6: $N_6 = 30$ teeth.

Determine the rotational velocity of gear 6 as gear 1 drives at 2000 rpm, ccw.

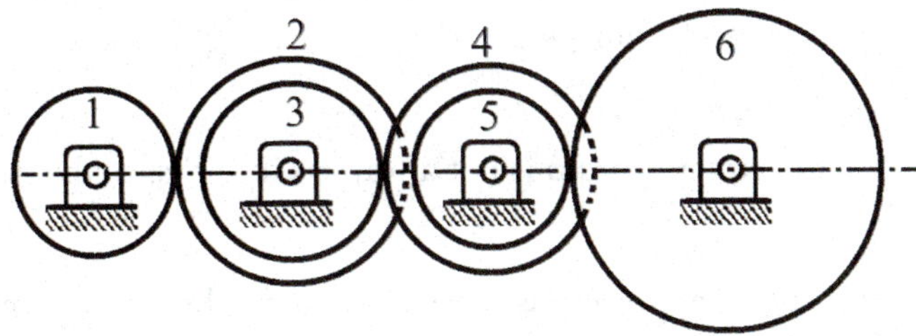

Figure 8.4: Gear train with fixed centers.

Solution:

For gear 1, $D_1 = N_1/P_d = 10/10$ in $= 1.0$ in.

For gears 3 and 4 having the same diametral pitch, one has

$$D_3 = \frac{N_3}{P_d} = \frac{15}{10}\, \text{in} = 1.50\, \text{in.}$$

Similarly, gears 5 and 6 have the same diametral pitch, therefore

$$D_6 = \frac{N_6}{P_d} = \frac{30}{10}\, \text{in} = 3.0\, \text{in.}$$

In turn, the train value becomes

$$\mu = \alpha_{12}\alpha_{34}\alpha_{56} = \left(\frac{-D_2}{D_1}\right)\left(\frac{-D_4}{D_3}\right)\left(\frac{-D_6}{D_5}\right)$$
$$= \left(\frac{-2.5}{1.0}\right)\left(\frac{-3.0}{1.5}\right)\left(\frac{-3.0}{1.5}\right) = -10.0.$$

Applying Equation (8.14), one has

$$\mu = \alpha_{16} = \frac{\omega_1}{\omega_6}. \quad \text{Hence,} \quad \omega_6 = \frac{\omega_1}{\mu} = \frac{2000\, \text{rpm}}{-10.0} = -200.0\, \text{rpm}$$

or the speed of the output gear is

$$\omega_6 = 200\,\text{rpm}, \qquad \text{clockwise.}$$

8.3.2   PLANETARY GEAR TRAINS

- Planetary gear trains can be used to achieve large speed reductions in a more compact space than in a conventional gear train with fixed centers.

- A greater benefit is the ability to readily change the train value.

- The gear train introduced in the last section has all the gear centers attached to a fixed frame of reference (ground). For planetary gear trains, gear centers can rotate with one another.

- The motion of a planetary gear train is not always as intuitive as that for fixed center gear trains.

- Planetary gear trains find applications in machine tools, hoists, automatic transmissions, automobile differentials, aircraft propeller reduction drives, and many others.

- A planetary gear train is also called an *epicyclic or cyclic gear train.*

Motion Analysis by Method of Superposition

The method consists essentially of the following three steps.

Step 1: Relax the constraint on the fixed link and temporarily assume that the carrier is fixed. Turn the previously fixed gear one revolution and calculate the effect on the entire train.

Step 2: Free all constraints and record the movement of rotating each link one revolution in the opposite direction to the original rotation in **Step 1**. Since this motion is combined with the motion in **Step 1**, the superimposed motion of the fixed gear equals zero.

Step 3: The motion of all other links are also determined by combining the rotations from the last two steps. Finally, velocities are proportional to the rotational movements.

Example 8.4

A planetary gear train is shown schematically in Figure 8.5. The carrier (link 2) serves as the input to the train. The sun gear (link 1) is the fixed gear and has 30 teeth. The planet gear (gear 3) has 35 teeth. The ring gear (link 4) serves as the output link from the train and has 120 teeth. Find the rotational velocity of the planet and ring gears of this gear train when the input shaft rotates at 1000 rpm clockwise.

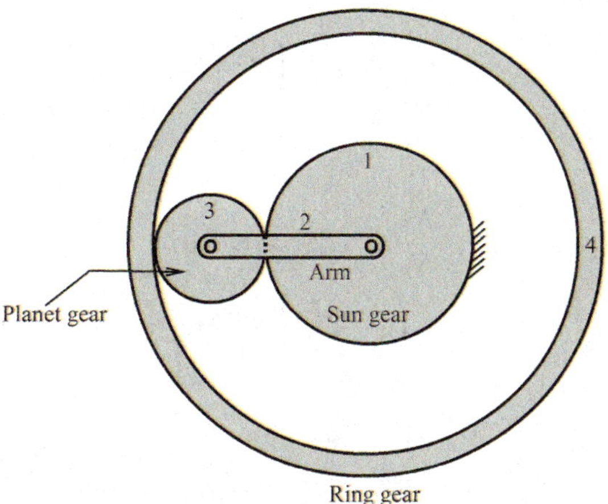

Figure 8.5: Planetary gear train.

Solution:

Step 1: Temporarily fix the carrier, then compute the motions of all gears as the previously fixed gear rotates one revolution. Thus:

Gear 1 rotates one revolution,

$$\Delta\theta_1 = +1 \text{ rev};$$

Gear 3 rotates

$$\Delta\theta_3 = -\left(\frac{30}{35}\right)\Delta\theta_1 = -0.857 \text{ rev}.$$

This is because the angular displacement is proportional to the angular velocity for a particular rotating gear, and since

$$\frac{\omega_3}{\omega_1} = \frac{R_1}{R_3}, \qquad \omega_1 R_1 = \omega_3 R_3.$$

But for the gear pair the diametral pitch is

$$P_d = \frac{N_3}{D_3} = \frac{N_1}{D_1}$$

such that

$$\frac{D_1}{D_3} = \frac{N_1}{N_3}, \quad \text{it leads to} \quad \frac{\omega_3}{\omega_1} = \frac{N_1}{N_3}.$$

Similarly, gear 4 rotates

$$\Delta\theta_4 = \left(\frac{30}{35}\right)\Delta\theta_3.$$

Therefore,

$$\frac{\omega_4}{\omega_3} = \frac{N_3}{N_4}, \quad \omega_4 R_4 = \omega_3 R_3.$$

$$\Delta\theta_4 = -\left(\frac{30}{35}\right)\left(\frac{35}{120}\right) 1\,\text{rev} = -0.25\,\text{rev}.$$

Step 2: In this step all links are rotated -1 revolution (that is, in the opposite direction to that in **Step 1**). This returns the sun gear to its original position, yielding a net movement of zero.

The method of superposition involves combining these two motions (that is, those in **Step 1** and **Step 2**), resulting in the actual planetary gear train motion. These are summarized in Table 8.3.

Table 8.3: Planetary gear train analysis

Link	Sun (gear 1), in rev	Planet (gear 3), in rev	Ring (gear 4), in rev	Carrier (gear 2), in rev
Step 1: rotate with fixed carrier	+1	-0.857	-0.25	0
Step 2: rotate all links	-1	-1	-1	-1
Step 3: total rotations	0	-1.857	-1.25	-1

The velocities can be determined by applying the ratios of rotations. Thus, the angular velocity of the sun, planet, and ring gears are, respectively,

$$\omega_s = \omega_1 = 0,$$

$$\omega_p = \omega_3 = -1.857\omega_c = -1.857\omega_2$$

$$= -1.857(-1000\,\text{rpm}) = 1857\,\text{rpm} \quad \text{cw,}$$

$$\omega_r = \omega_4 = -1.25\omega_c = -1.25\omega_2$$

$$= -1.25(-1000\,\text{rpm}) = 1250\,\text{rpm} \quad \text{cw.}$$

Motion Analysis by Method of Formula

This method can be applied to planetary gear trains.

With reference to the planetary gear train shown in Figure 8.6, one has

$$\omega_{24} = \omega_2 - \omega_4 = \omega_{21} - \omega_{41},$$

$$\omega_{34} = \omega_3 - \omega_4 = \omega_{31} - \omega_{41},$$

and dividing the first by the second equation gives

$$\frac{\omega_{24}}{\omega_{34}} = \frac{\omega_{21} - \omega_{41}}{\omega_{31} - \omega_{41}}.$$

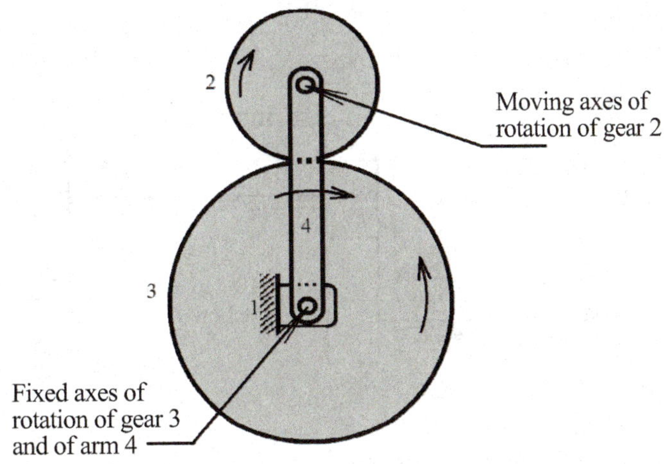

Figure 8.6: Planetary gear train.

In Figure 8.6 if gear 3 is considered the first gear and gear 2 the last gear, then the last equation can be written as

$$\frac{\omega_{LA}}{\omega_{FA}} = \frac{\omega_L - \omega_A}{\omega_L - \omega_A}, \tag{8.15}$$

where

$$\omega_L = \quad \text{angular velocity of last gear in train relative to fixed link,}$$

$$\omega_F = \quad \text{angular velocity of first gear in train relative to fixed link,}$$

$$\omega_A = \quad \text{angular velocity of arm relative to fixed link, and}$$

$$\frac{\omega_{LA}}{\omega_{FA}} = \quad \text{velocity ratio of last gear to first gear, both relative to arm.}$$

- In applying Equation (8.15) the first gear and the last gear must be gears that mesh with the gear or gears having planetary motion.

- The first and last gears must be on parallel shafts as angular velocities can not be dealt with algebraically unless the vectors representing these velocities are parallel.

- Sometimes, a planetary gear train can not be dealt with by a single application of Equation (8.15) and therefore, a second application of Equation (8.15) is required. This remark will be illustrated in the following two examples.

Example 8.5

Consider the gear system shown in Figure 8.7. Arm 6 and gear 5 are driven clockwise at 200 and 60 rad/s, respectively, viewed from the left-hand end. Determine the angular velocity of gear 2, ω_{21} and its direction.

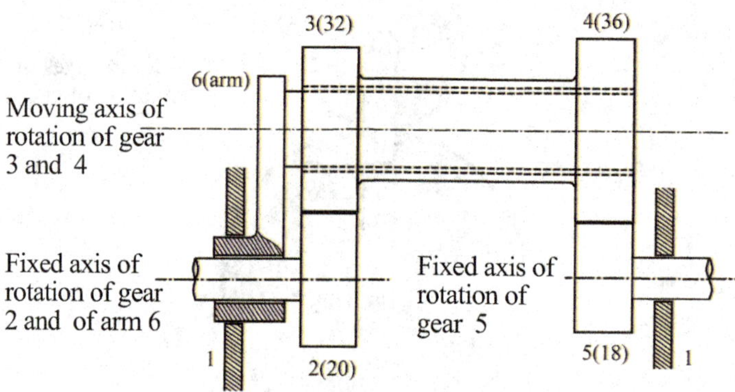

Figure 8.7: Section view of planetary gear train.

Solution:

Since gears 5 and 2 undergo planetary motion, and they are on parallel shafts hence Equation (8.15) can be applied to this example. The velocity of gear 5 is given and that of gear 2 is required. Therefore, one can assume gear 5 is the first gear and gear 2 is the last gear.

Applying Equation (8.15),

$$\frac{\omega_{LA}}{\omega_{FA}} = \frac{\omega_L - \omega_A}{\omega_L - \omega_A},$$

and with the aforementioned identification for the system, one has

$$\frac{\omega_{26}}{\omega_{56}} = \frac{\omega_{21} - \omega_{61}}{\omega_{51} - \omega_{61}}.$$

Recalling from Equation (8.14) that the rotational velocity is inversely proportional to radius of the pitch circle while the radius of pitch circle is proportional to the number of teeth of the gear, therefore the lhs of the last equation becomes

$$\frac{\omega_{26}}{\omega_{56}} = \left(-\frac{\omega_2}{\omega_3}\right)\left(-\frac{\omega_4}{\omega_5}\right) = \left(\frac{N_3}{N_2}\right)\left(\frac{N_5}{N_4}\right) = \frac{(32)(18)}{(20)(36)} = \frac{4}{5},$$

since gear 2 meshes gear 3, and gear 4 meshes gear 5. Thus,

$$\frac{4}{5} = \frac{\omega_{21} - 200}{60 - 200}.$$

This gives

$$\omega_{21} = 128.0 \, \text{rad/s}.$$

Since the sign of ω_{21} is the same as that of ω_{51} and ω_{61}, therefore ω_{21} is in the clockwise direction, viewed from the left-hand end.

Example 8.6
If a fixed internal gear 7 is added to the train in Figure 8.7 to mesh with gear 4, as shown in Figure 8.8, determine the angular velocity of gear 5, ω_{51} and its direction of rotation, given that now gear 2 rotates clockwise (viewed from the left-hand end) with an angular velocity, $\omega_{21} = 50$ rad/s.

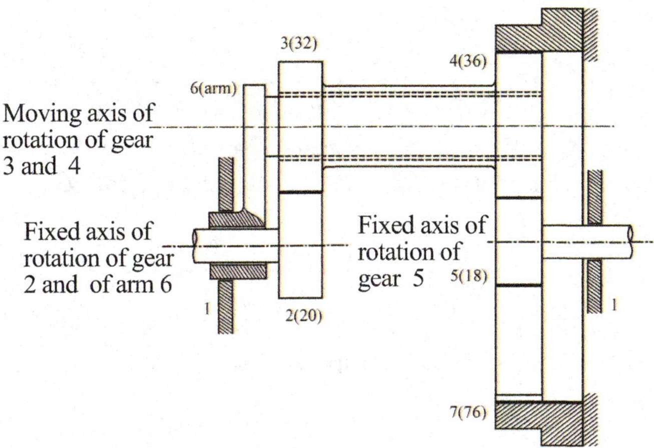

Figure 8.8: Section view of planetary gear train with a fixed ring gear.

Solution:

In this example, Equation (8.15) will be applied twice. The first application involves with *gears 2, 3, 4, and 5, and arm 6*. In this case, gear 2 is the first gear and gear 5 is the last gear so that upon application of Equation (8.15) one has

$$\frac{\omega_{56}}{\omega_{26}} = \frac{\omega_{51} - \omega_{61}}{\omega_{21} - \omega_{61}}.$$

Again, recalling from Equation (8.14) that the rotational velocity is inversely proportional to radius of the pitch circle while the radius of pitch circle is proportional to the number of teeth of the gear, therefore the lhs of the last equation becomes

$$\frac{\omega_{56}}{\omega_{26}} = \left(-\frac{\omega_5}{\omega_4}\right)\left(-\frac{\omega_3}{\omega_2}\right) = \left(\frac{N_4}{N_5}\right)\left(\frac{N_2}{N_3}\right) = \frac{(36)(20)}{(18)(32)} = \frac{5}{4},$$

since gear 5 meshes gear 4, and gear 2 meshes gear 3. Clearly, one can make use of the corresponding result of the last example by recognizing that the present result is the reciprocal to that in the last example. Applying the above result, one has

$$\frac{5}{4} = \frac{\omega_{51} - \omega_{61}}{\omega_{21} - \omega_{61}} = \frac{\omega_{51} - \omega_{61}}{50 - \omega_{61}}. \tag{8.16}$$

Since there are two unknowns in this equation so that a second application of Equation (8.15) is required in order to provide another equation. In this second application, *gears 2, 3, 4, and 7, and arm 6* are considered. Thus, by Equation (8.15), one obtains

$$\frac{\omega_{76}}{\omega_{26}} = \frac{\omega_{71} - \omega_{61}}{\omega_{21} - \omega_{61}}.$$

The lhs of the last equation becomes

$$\frac{\omega_{76}}{\omega_{26}} = \left(\frac{\omega_7}{\omega_4}\right)\left(-\frac{\omega_3}{\omega_2}\right) = \left(\frac{N_4}{N_7}\right)\left(-\frac{N_2}{N_3}\right) = \frac{-(36)(20)}{(76)(32)} = -\frac{9(5)}{36(4)},$$

since gear 7 meshes gear 4, and gear 2 meshes gear 3. Note that there is no sign change for gears 4 and 7 since they are internal gears while there is a sign change for gears 2 and 3 because they are external gears. Thus,

$$-\frac{9(5)}{36(4)} = \frac{\omega_{71} - \omega_{61}}{\omega_{21} - \omega_{61}} = \frac{0 - \omega_{61}}{50 - \omega_{61}} \tag{8.17}$$

in which $\omega_{71} = 0$ as gear 7 is fixed.

Solving for ω_{61} from Equation (8.17) above gives

$$\omega_{61} = -22.73 \text{ rad/s}.$$

Substituting this result into Equation (8.16), one obtains

$$\omega_{51} = 68.18 \text{ rad/s},$$

with the direction of rotation same as that of ω_{21}.

8.4   APPLICATION OF GEAR TRAINS IN AUTOMOTIVE TRANSMISSION ANALYSIS

This application is essentially a problem from [1] concerning a conventional automotive transmission system. It is shown diagrammatically in Figure 8.9.

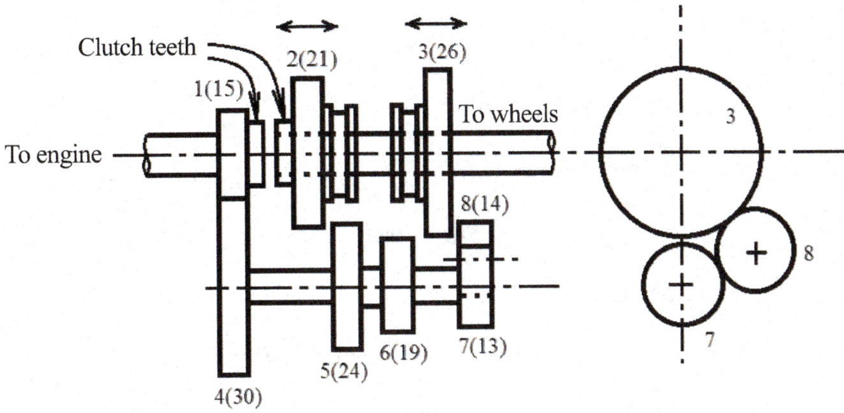

Figure 8.9: Section view of an automotive gear train.

The transmission of power is as follows.

Lower gear, gear 3 shifted to mesh with gear 6. Transmission of power is through gears 1, 4, 6, and 3.

Second gear, gear 2 shifted to mesh with gear 5. Transmission of power is through gears 1, 4, 5 and 2.

High gear, gear 2 shifted so that clutch teeth on end of gear 2 mesh with clutch teeth on end of gear 1. Direct drive results.

Reverse gear, gear 3 shifted to mesh with gear 8. Transmission of power is through gears 1, 4, 7, 8, and 3.

A car equipped with this transmission has a differential ratio of 2.8:1 and a tire outside diameter of 720 mm.

For illustration, the computation of the engine speed of the car under the given condition: Low gear and car traveling at 16 km/h, is considered in Figure 8.9.

Step 1: Relate the engine speed to car velocity from the condition specified above. Since

$$v = R\omega_w,$$

where

$$\omega_w = v/R, \text{ wheel rotational speed in rad/s,}$$
$$v = \text{car speed in m/s, and}$$
$$R = \text{wheel radius in m} = 0.720/2 \text{ m} = 0.360 \text{ m.}$$

Converting units on v:

$$v \text{ m/s} = v' \ \ (\text{km/h})(1000 \text{ m/km}) \ (\text{h}/3600 \text{ s}), \text{ where } \ \ v' \ \ \text{ is in km/h.}$$

Thus, $v = v'/3.6$ and $\omega_w = v'/(3.6R)$.

Step 2: Relate ω_w to enginee speed ω_E,

$$\omega_E = \omega_w r_o r_i,$$

where $r_o = $ differential ratio $= 2.8$, given

$$r_i = \text{transmission ratio, in which} \ \ i = 1, 2, 3, 4.$$

At low gear, that is, $i = 1$,

$$r_1 = \alpha_{13} = \frac{\omega_1}{\omega_3} = \left(-\frac{\omega_1}{\omega_4}\right)\left(-\frac{\omega_6}{\omega_3}\right) = \left(\frac{N_4}{N_1}\right)\left(\frac{N_3}{N_6}\right)$$

in which the negative signs are due to the fact that gears 1 and 4, and gears 3 and 6 are both external. Therefore,

$$r_1 = \left(\frac{N_4}{N_1}\right)\left(\frac{N_3}{N_6}\right) = \left(\frac{30}{15}\right)\left(\frac{26}{19}\right) = \frac{52}{19}.$$

For $v' = 16$ km/h,

$$\omega_E = \frac{16}{3.6}(2.8)\left(\frac{52}{19}\right)\frac{1}{0.36} \text{ rad/s.}$$

Therefore, the speed of the engine at low gear is

$$\omega_E = 94.61 \text{ rad/s} \quad \text{or} \quad 903.43 \text{ rpm.}$$

8.5   EXERCISES

8.1.   If the radii of a pinion and gear are increased so that each becomes a rack, the length of action theoretically becomes a minimum. Determine the equation for the length of action under these conditions and calculate the maximum contact ratio for $22\frac{1}{2}^{\circ}$ full-depth system.

8.2. For a pair of standard spur gears in mesh, determine:

(a) an equation for the center distance C as a function of the numbers of teeth and diametral pitch, and

(b) the various combinations of 25° full-depth gears that can be used to operate at a center distance of 6.00 in with an angular velocity ratio of 5:1. For the foregoing solution the diametral pitch is kept below 16.

8.3. For a pressure angle of 20° in the full-depth system, calculate the minimum number of teeth in a pinion to mesh with a rack without involute interference. In addition, calculate the number of teeth in a pinion to mesh with a gear of equal size (which means $N_1 = N_2 = N$) without involute interference.

8.4. A 120-pitch (that is, the diametral pitch), 20° full-depth pinion of 40 teeth drives a gear of 80 teeth. Calculate the contact ratio.

8.5. A 6-pitch, 20° full-depth pinion with 40 teeth drives a rack. Calculate the length of action and the contact ratio.

8.6. In the planetary gear train shown in Figure 8.10, gear 2 turns at 500 rpm in the direction indicated. Determine the speed and direction of rotation of arm 6 if gear 5 rotates at 400 rpm in the same direction as gear 2.

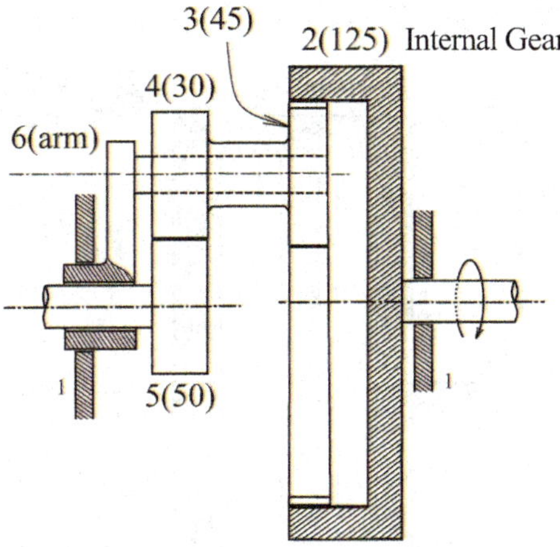

Figure 8.10: Section view of planetary gear train with a rotating ring gear.

8.7. For the gear train shown in Figure 8.11, shaft A rotates at 200 rpm and shaft B at 500 rpm in the direction shown. Determine the speed and direction of rotation of shaft C.

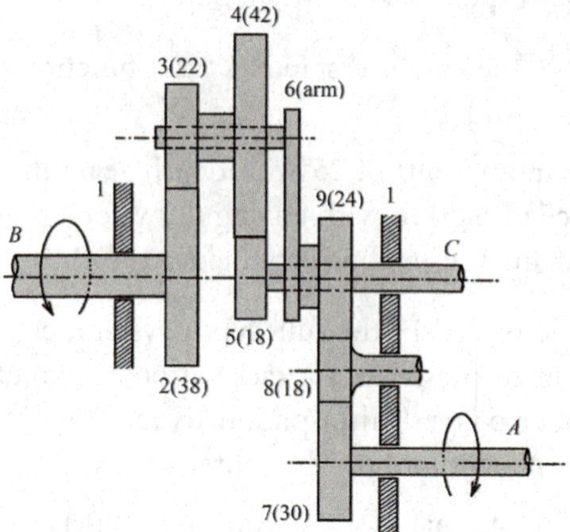

Figure 8.11: Section view of gear train with two inputs.

8.8. A planetary gear train for a large reduction is shown in Figure 8.12. If shaft A is connected to the motor and it is rotating with a constant angular velocity of 1000 rpm, ccw viewing from right to left of the figure, determine the angular velocity ratio of $\frac{\omega_A}{\omega_B}$ by applying the formula method. Calculate the angular velocity of shaft B.

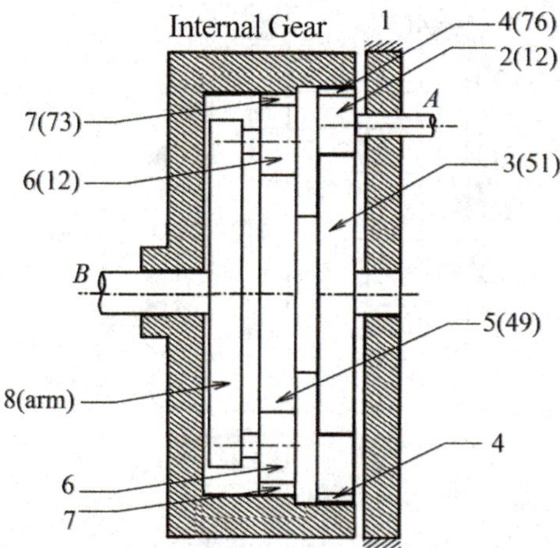

Figure 8.12: Section view of planetary gear train for large reduction.

REFERENCES

[1] Mabie, H. H. and Ocvirk, F. W., *Mechanisms and Dynamics of Machinery*, 3rd edition-SI Version, New York, John Wiley & Sons, 1978. 137, 142, 155

[2] Spiegel, M. R., *Mathematical Handbook of Formulas and Tables*, Schaum's Outline Series in Mathematics, New York, McGraw-Hill, 1968. 159

8A   APPENDIX: PROPERTIES OF INVOLUTE

The involute is a curve described by the end point P of a taut string when it unwinds from a circle of radius R [2]. This curve is shown in Figure 8A.1 and can be expressed in complex number form as $z = Re^{i\theta}(1 - i\theta)$. The proof of this equation is given in the following while the objective of this appendix is to provide the parametric equations for the involute.

Before unwrapping, the tip of the string is at point A on the circumference of center O. When the tip of the taut string is unwrapped to the new position P one has

$$BP = AB = R\theta, \tag{8A.1}$$

where B is the tangent point on the circumference of the circle.

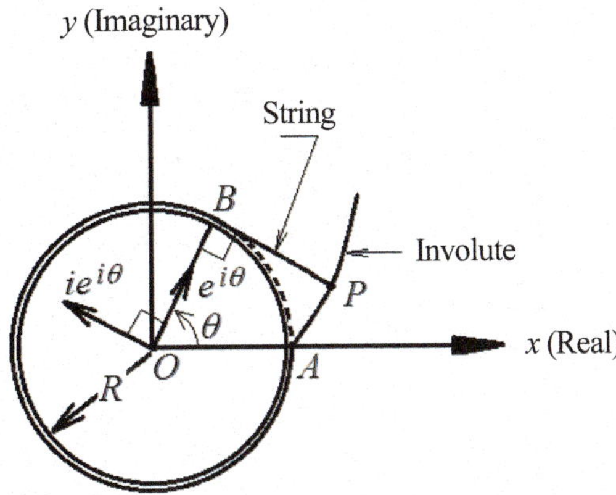

Figure 8A.1: Formation of involute.

The position vector

$$\overrightarrow{OP} = z = \overrightarrow{OB} + \overrightarrow{BP} \tag{8A.2}$$

in which $\overrightarrow{OB} = Re^{i\theta}$, $\overrightarrow{BP} = R\theta(-ie^{i\theta})$ since from Equation (8A.1) $R\theta$ is the magnitude of $\overrightarrow{BP}$ whose direction is in opposite to that of $ie^{i\theta}$ (Recall, from Chapter 5, that multiplying the vector by the imaginary number i is equivalent to rotating the vector by 90° ccw). Thus, Equation (8A.2) becomes

$$z = Re^{i\theta}(1 - i\theta). \tag{8A.3}$$

Equation (8A.3) is the expression for an involute. It gives the location of point P for a specified angular displacement θ.

Expanding Equation (8A.3),

$$\begin{aligned}
z = x + iy &= R(\cos\theta + i\sin\theta)[1 - i\theta] \\
&= R(\cos\theta + i\sin\theta) - i\theta R(\cos\theta + i\sin\theta) \\
&= R\cos\theta + iR\sin\theta - i\theta R\cos\theta + \theta R\sin\theta.
\end{aligned}$$

Equating real and imaginary parts, one obtains

$$x = R(\cos\theta + \theta\sin\theta), \qquad y = R(\sin\theta - \theta\cos\theta). \tag{8A.4}$$

These are the parametric equations of the involute and can be applied to generate the involute tooth form for specific angular displacement θ in the xy-plane.

CHAPTER 9

Analysis of Dynamic Forces in Machinery

It is understood that analysis of dynamic forces in machinery encompasses dynamic balance and vibration in rotating machines and engines, in addition to contact dynamics in machine components such as bearings. Everyone of these important branches of dynamic force analysis requires a treatment well beyond the scope of the present book. To limit the scope, therefore in this chapter only the D'Alembert's principle which is included in Section 9.1, method of virtual work which is outlined in Section 9.2, gyroscopic forces in Section 9.3, and representative questions and solutions are presented in Section 9.4.

9.1 D'ALEMBERT'S PRINCIPLE AND INERTIA FORCE

Starting from Newton's law of motion,

$$\vec{R} = m\vec{a}_G, \tag{9.1}$$

where $\vec{R}, m$, and $\vec{a}_G$ are the force, mass, and acceleration through the center of gravity G of the body. Equation (9.1) can be rewritten as

$$\vec{R} + \vec{F}^i = 0 \tag{9.2}$$

in which $\vec{F}^i = -m\vec{a}_G$ is the apparent force.

Similarly, for the rotating counterpart the applied and apparent torques of the body about an axis through the point O can be written as

$$\vec{T}_o + \vec{T}^i_o = 0, \tag{9.3}$$

where $\vec{T}_o$ and $\vec{T}^i_o = -I_o\vec{\alpha}$ are, respectively, the applied and apparent torques of the body about an axis through the point O. I_o is the mass moment of inertia of the body, and $\vec{\alpha}$ is the angular acceleration about an axis through the point O.

The external forces acting on a rigid body are equivalent to the effective forces of the various particles forming the body. This is referred to as the D'Alembert's principle. Equations (9.1) and (9.2) describe the body in equilibrium under the combined action of external force system (which, in the case of fixed axis of rotation, includes the bearing reactions) and the apparent force $\vec{F}^i$.

9.2   ANALYSIS OF MECHANISMS BY VIRTUAL WORK METHOD

The principle of virtual work applies to systems in equilibrium. In the virtual work method one does not consider constraint forces because work done by virtual displacement is assumed zero.

A virtual displacement $\delta \vec{s}_j$ of the system is a consequence of assumed infinitesimal change of system co-ordinates with the time being held constant. The reason why it is referred as virtual rather than real is because no actual displacement can take place without the passage of time.

If a physical system is in equilibrium and it undergoes an infinitesimally small displacement, compatible with the constraints, the net work done by the forces during the process is zero. Symbolically,

$$\delta W = \sum_{j=1}^{n} \vec{F}_j \bullet \delta \vec{s}_j + \sum_{j=1}^{n} \vec{T}_j \bullet \delta \vec{\theta}_j = 0, \tag{9.4}$$

where the integer n is the number of forces in the system.

Equation (9.4) may be modified for dynamic cases such that

$$\sum_{j=1}^{n} \vec{F}_j \bullet \frac{\delta \vec{s}_j}{\delta t} + \sum_{j=1}^{n} \vec{T}_j \bullet \frac{\delta \vec{\theta}_j}{\delta t} = 0. \tag{9.5}$$

This is permissible for every virtual displacement taking place in the same interval of time such that as $\delta t \to 0$ Equation (9.5) can be written as

$$\sum_{j=1}^{n} \vec{F}_j \bullet \frac{d \vec{s}_j}{dt} + \sum_{j=1}^{n} \vec{T}_j \bullet \frac{d \vec{\theta}_j}{dt} = 0,$$

or

$$\sum_{j=1}^{n} \vec{F}_j \bullet \vec{v}_j + \sum_{j=1}^{n} \vec{T}_j \bullet \vec{\omega}_j = 0. \tag{9.6}$$

This equation is particularly useful in the dynamic force analysis of machinery.

9.3   GYROSCOPIC FORCES

In machines with rotating components or elements of high moment of inertia gyroscopic forces are induced when such components or elements undergo change of direction of motion.

From engineering dynamics [1] it is known that the rate of change of angular momentum $\vec{H}$ of a body with respect to time is equal to an applied torque $\vec{T}$. That is, one can write

$$\vec{T} = d\vec{H}/dt \tag{9.7}$$

in which the angular momentum is defined by $\vec{H} = I\vec{\omega}_s$ with $\vec{\omega}_s$ being the constant angular velocity of the spinning body about an axis through its mass center, and I the moment of inertia of the body about its spinning axis. The direction and axis of spin is illustrated in Figure 9.1. The spin axis is considered fixed in the latter figure.

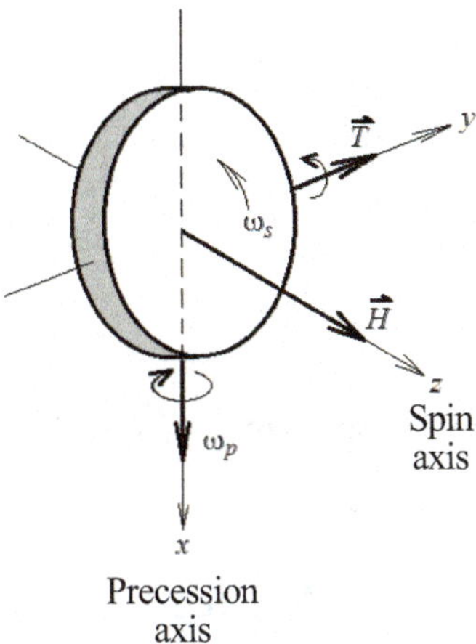

Figure 9.1: Direction and axis of spin of a body.

If the spin axis changes its angular position due to, say, the moving body traversing a curved path in plane motion, as shown in Figure 9.2a, gyroscopic force results. For constant $\vec{\omega}_s$ the magnitude of the angular momentum remains constant for an infinitesimally small angular displacement $\Delta\theta$ of the spinning axis as shown in the vector diagram in Figure 9.2b. With reference to the latter figure, the magnitude of $\Delta\vec{H}$ is given by

$$\Delta H = (I\omega_s)\Delta\theta.$$

Therefore, the rate of change of H with respect to time t is given by

$$\lim_{\Delta t \to 0} \frac{\Delta H}{\Delta t} = \lim_{\Delta t \to 0} (I\omega_s)\frac{\Delta\theta}{\Delta t} \quad \text{or} \quad \frac{dH}{dt} = (I\omega_s)\frac{d\theta}{dt}$$

in which $\frac{d\theta}{dt}$ is the angular velocity of precession of the spin axis and is generally denoted by ω_p. Thus, the last equation can be written as

$$\frac{dH}{dt} = T = I\omega_s\omega_p \quad \text{or in vector form} \quad \vec{T} = \vec{\omega}_p \times (I\vec{\omega}_s). \tag{9.8}$$

This equation is illustrated in Figure 9.2c in which $\vec{T}$ is a couple and is called the gyroscopic couple.

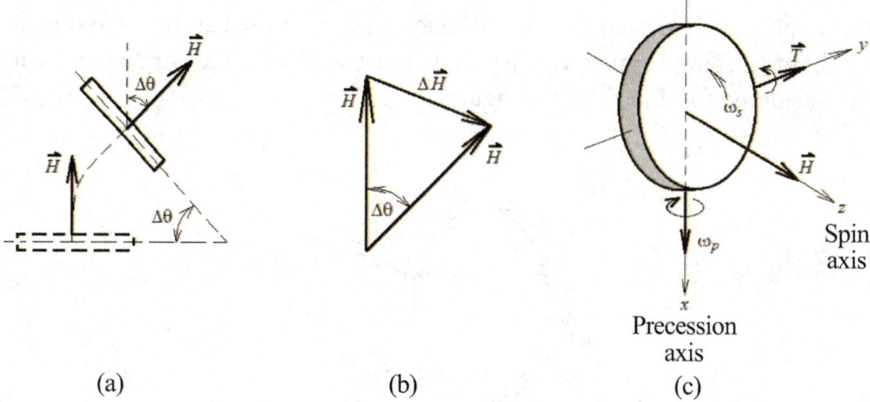

(a)   (b)   (c)

Figure 9.2: Gyroscopic force in a rotor: (a) change of spin axis, (b) change in H, and (c) spin, precession, and torque.

9.4   QUESTIONS AND SOLUTIONS

In this section, Equations (9.6) and (9.8) are applied to the solution of three representative machine systems. The first is concerned with the determination of the crankshaft torque necessary to hold the linkage in equilibrium by using Equation (9.6). Applying the latter equation, the second question deals with the determination of shaft torque required to hold a cylinder of the donkey engine in equilibrium. The main difference between the first and second questions is that in the latter question the velocity diagram is constructed from the information given in the acceleration diagram. The third question is concerned with the application of Equation (9.8) for the analysis of the effect of gyroscopic forces on the bearing forces of a jet airplane.

Example 9.1

Applying the virtual work method, calculate the crankshaft torque $\vec{T}_2$ necessary to hold the linkage in equilibrium for the slider-crank mechanism shown in Figure 9.3a. Note that the slider is supposed to be operating in the cylinder due to the compressed gas inside. However, the gas force inside the cylinder may be assumed zero. Required parameters and operating condition are provided below:

$$Ag_3 = 50.8 \text{ mm}, \qquad \text{stroke} = 2(AO_2) = 102 \text{ mm},$$
$$\text{connecting rod having } I_3 = 0.0080 \text{ kg m}^2, m_3 = 0.90 \text{ kg}, AB = 204 \text{ mm},$$
$$\text{piston mass } m_4 = 2.0 \text{ kg, and constant } \omega_2 = 2000 \text{ rpm, ccw.}$$

Solution:
The solution of this question consists of four parts:

(a) construction of the configuration diagram,

(b) construction of the velocity diagram,

(c) construction of the acceleration diagram, and

(d) application of the principle of virtual work (virtual work method).

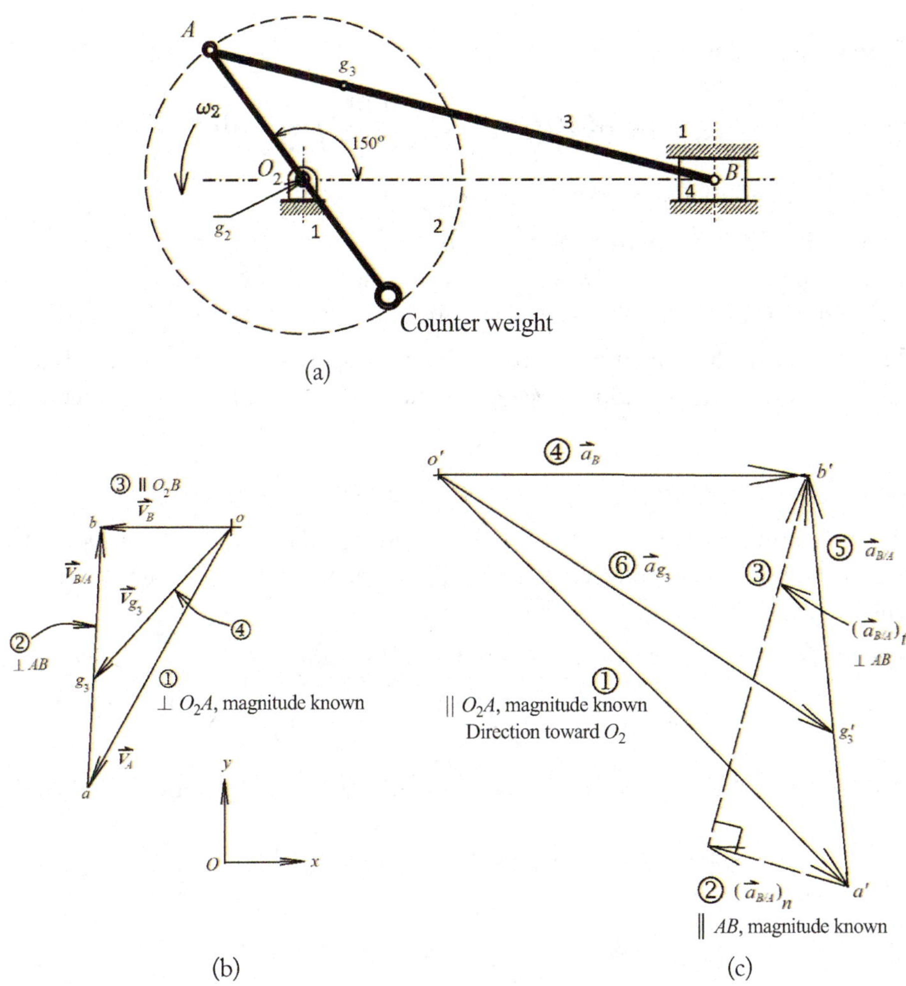

Figure 9.3: Slider-crank mechanism: (a) mechanism and configuration diagram, (b) velocity diagram, and (c) acceleration diagram.

(a) *Construction of configuration diagram*

With the given parameters, choose a scale so that the configuration diagram can be constructed as in Figure 9.3a.

(b) *Construction of velocity diagram*

The angular velocity of the crank is

$$\omega_2 = \frac{2\pi (2000)}{60} \frac{\text{rad}}{\text{s}} = 209.5 \frac{\text{rad}}{\text{s}}, \text{ ccw.}$$

Magnitude of velocity at point A,

$$v_A = \omega_2(AO_2) = 209.5 \left(\frac{0.102}{2}\right) \frac{\text{m}}{\text{s}} = 10.68 \frac{\text{m}}{\text{s}},$$

and direction of velocity is perpendicular to AO_2.

Magnitude of relative velocity $\vec{v}_{B/A}$ is unknown but it is perpendicular to AB which is link 3. The line of this relative velocity ab intersects with that of velocity at point B as shown in Figure 9.3b in which a scale has been chosen.

The point g_3 in the velocity diagram is determined by the *image method*. That is, in the scaled construction, the ratios of velocities and that of lengths are equal such that

$$\frac{ag_3}{bg_3} = \frac{Ag_3}{Bg_3}.$$

Complete joining the line between points o and g_3 then og_3 is the velocity $\vec{v}_{g_3}$. Note that in the velocity diagram circled integer designates the step number in the graphical construction.

From the velocity diagram, the magnitudes of the velocities

$$v_B = 4.15 \frac{\text{m}}{\text{s}}, \quad v_{B/A} = 9.3 \frac{\text{m}}{\text{s}}, \quad v_{g_3} = 8.6 \frac{\text{m}}{\text{s}}.$$

With the co-ordinates chosen as shown in Figure 9.3b, the velocities

$$\vec{v}_B = \vec{v}_{g_4} = -4.15 \vec{i} \ \frac{\text{m}}{\text{s}}, \quad \vec{v}_{g_3} = -5.05 \vec{i} - 6.95 \vec{j} \ \frac{\text{m}}{\text{s}}.$$

The angular velocities

$$\vec{\omega}_2 = 209.5 \vec{k} \ \frac{\text{rad}}{\text{s}},$$

$$\vec{\omega}_3 = \frac{v_{B/A}}{BA} \vec{k} \ \frac{\text{rad}}{\text{s}} = \frac{9.3}{0.204} \vec{k} \ \frac{\text{rad}}{\text{s}} = 45.6 \vec{k} \ \frac{\text{rad}}{\text{s}},$$

$$\vec{\omega}_4 = 0.$$

(c) *Construction of acceleration diagram*

First, choose a scale for the construction. Starting from the acceleration at point A, since ω_2 is constant and therefore angular acceleration $\alpha_2 = 0$ and hence the acceleration at point A consists of its normal component only. Its magnitude is

$$a_A = (a_A)_n = \omega_2^2(AO_2) = 2238.5 \ \frac{m}{s^2}.$$

The normal component of the relative acceleration

$$(a_{B/A})_n = (BA)\omega_3^2 = \frac{v_{B/A}^2}{BA} = \frac{9.3^2}{0.204} \ \frac{m}{s^2} = 423.97 \ \frac{m}{s^2}$$

whose direction is from point B directed toward point A.

The magnitude of the tangential component of $\vec{a}_{B/A}$ is unknown but its direction is perpendicular to the normal component or it is parallel to the relative velocity $\vec{v}_{B/A}$. The line of this tangential component intersects with the acceleration $\vec{a}_B$ at point b' of the acceleration diagram shown in Figure 9.3c.

The point g_3 in the acceleration diagram is determined by the *image method*. Thus, in the scaled construction the ratios of acceleration and that of the lengths are equal such that

$$\frac{a'g_3}{b'g_3} = \frac{Ag_3}{Bg_3}.$$

From the acceleration diagram, the magnitudes of the accelerations

$$a_B = 1652.5 \ \frac{m}{s^2}, \ (a_{B/A})_t = 1070 \ \frac{m}{s^2}, \ a_{B/A} = 1155 \ \frac{m}{s^2}, \ a_{g_3} = 2050 \ \frac{m}{s^2}.$$

With reference to the co-ordinates chosen, the accelerations are

$$\vec{a}_B = \vec{a}_{g_4} = 1652.5\vec{i} \ m/s^2, \quad \vec{a}_{g_3} = 1870\vec{i} - 840\vec{j} \ m/s^2.$$

The angular accelerations are

$$\vec{\alpha}_3 = \frac{(a_{B/A})_t}{BA}\vec{k} \ \frac{rad}{s^2} = \frac{1070}{0.204}\vec{k} \ \frac{rad}{s^2} = 5245\vec{k} \ \frac{rad}{s}, \quad \vec{\omega}_4 = 0.$$

(d) *Application of virtual work method*

Recall that $\vec{i} \bullet \vec{j} = 0$, $\vec{j} \bullet \vec{j} = 1$, and so on. Applying Equation (9.6), one has

$$\vec{F}_{03} \bullet \vec{v}_{g_3} + \vec{F}_{04} \bullet \vec{v}_{g_4} + \vec{T}_2 \bullet \vec{\omega}_2 + \vec{T}_{03} \bullet \vec{\omega}_3 + \vec{T}_{04} \bullet \vec{\omega}_4 = 0. \qquad (9.9)$$

Now, consider every term in this equation in which

$$\vec{F}_{03} \bullet \vec{v}_{g3} = -m_3 \vec{a}_{g3} \bullet \vec{v}_{g3}$$
$$= -(0.90)(1870\,\vec{\imath} - 840\,\vec{\jmath}) \bullet (-5.05\,\vec{\imath} - 6.95\,\vec{\jmath})\ \text{Nm/s}$$
$$= 3244.95\ \text{Nm/s}.$$
$$\vec{F}_{04} \bullet \vec{v}_{g4} = -m_4 \vec{a}_{g4} \bullet \vec{v}_{g4}$$
$$= -(2.0)(1652.5\,\vec{\imath}) \bullet (-4.15\,\vec{\imath})\ \text{Nm/s}$$
$$= 13715.75\ \text{Nm/s}.$$
$$\vec{T}_2 \bullet \vec{\omega}_2 = (T_2 \vec{k}) \bullet (209.5\,\vec{k}) = 209.5 T_2.$$
$$\vec{T}_{03} \bullet \vec{\omega}_3 = -I_3 \vec{\alpha}_3 \bullet \vec{\omega}_3 = -0.0080(5245\,\vec{k}) \bullet (45.6\,\vec{k})\ \text{Nm/s}$$
$$= -1913.376\ \text{Nm/s}.$$
$$\vec{T}_{04} \bullet \vec{\omega}_4 = 0 \quad \text{since} \quad \vec{\omega}_4 = 0.$$

Substituting all the above results into Equation (9.9), one obtains

$$3244.95 + 13715.75 + 209.5 T_2 - 1913.376\ \text{Nm/s} = 0.$$

This gives $T_2 = -71.82$ Nm/s or $T_2 = 71.82$ Nm/s, cw.

Example 9.2

By applying the virtual work method, calculate the shaft torque $\vec{T}_2$ which must be applied to link 2 of the donkey engine in Figure 9.4a to hold the mechanism in equilibrium. The system parameters are:

$$O_2 A = 102\,\text{mm}, \ m_2 = 4.53\,\text{kg}, \ I_2 = 0.0114\,\text{kg m}^2,$$
$$AB = 356\,\text{mm}, \ m_3 = 6.80\,\text{kg}, \ I_3 = 0.068\,\text{kg m}^2,$$
$$Ag_3 = 254\,\text{mm}, \ m_4 = 11.30\,\text{kg}, \ I_4 = 0.1080\,\text{kg m}^2.$$

This question is similar to that in [2] except for the differences in steam pressure and piston area.

Solution:
The solution of this question consists of four parts:

(a) construction of the configuration diagram,

(b) construction of the velocity diagram,

(c) making use of the given acceleration diagram, and

(d) application of the principle of virtual work (virtual work method).

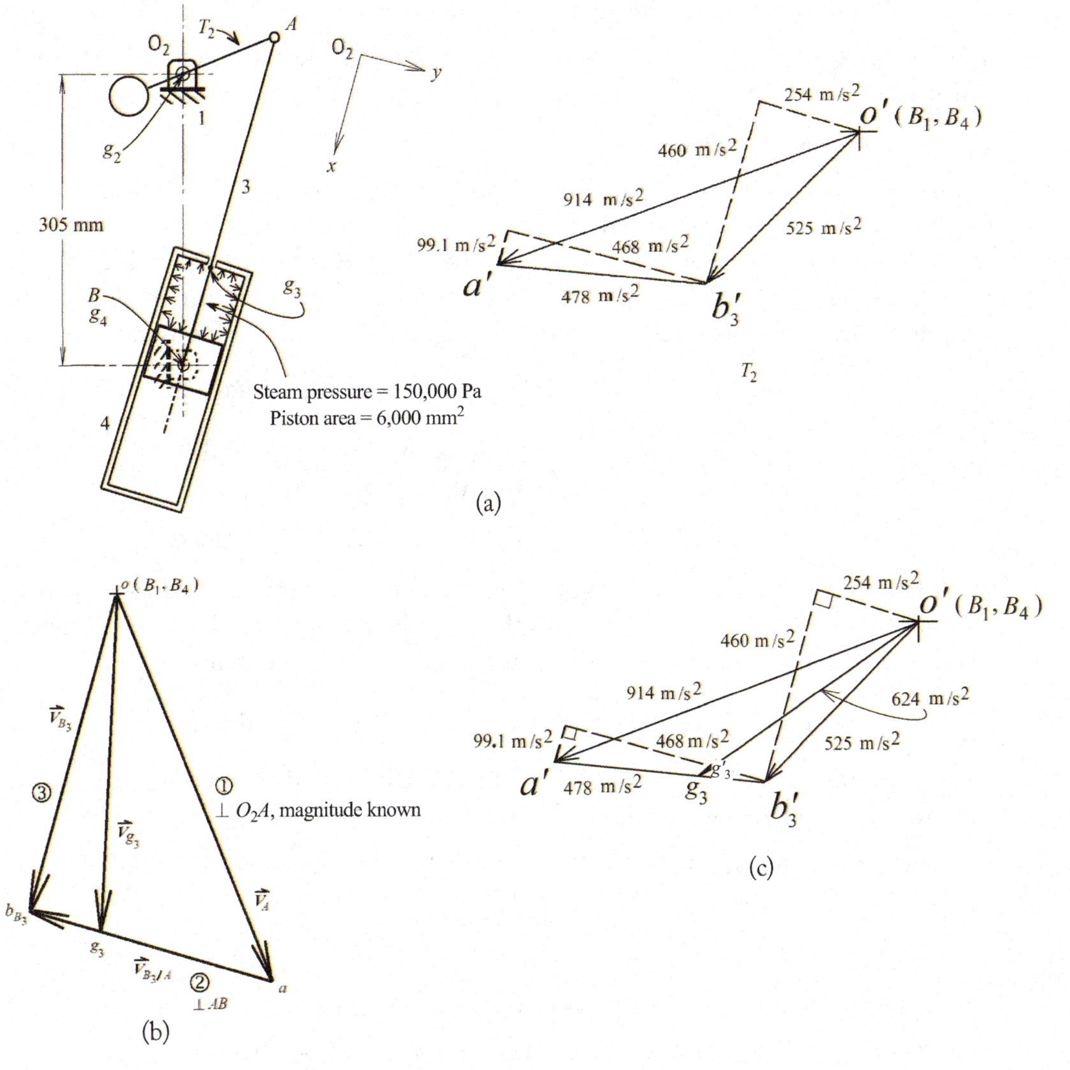

Figure 9.4: Donkey engine: (a) mechanism with acceleration diagram. (b) velocity diagram, and (c) image method for $\vec{a}_{g_3}$.

Before the above parts are considered one requires to find the steam force $\vec{F}_s$. This is the product of the piston area and steam pressure. Since the steam pressure $p_s = 150,000$ N/m^2 and the piston area $A_p = 6,000$ mm^2, therefore the magnitude of $\vec{F}_s$ is $F_s = 900$ N whose direction is parallel to the chosen x-axis as shown in Figure 9.4a.

(a) *Construction of configuration diagram*

With the given parameters, a scale of 1:5 was chosen and the configuration diagram was constructed as in Figure 9.4a.

(b) *Construction of velocity diagram*

In this part, the velocities of moving links are required and therefore the given acceleration diagram in Figure 9.4a is employed. With reference to the latter figure, the acceleration at A is $a'o'$ which is parallel to AO_2 and toward the center of curvature O_2. This means that there is no tangential component of the acceleration at A. This, in turn, implies that the angular acceleration of link 2 is zero. Thus, the angular velocity $\vec{\omega}_2$ is a constant. Therefore,

$$a_A = 914 \,\text{m/s}^2 = (a_A)_n = \omega_2^2(AO_2) = v_A^2/(AO_2).$$

This gives

$$v_A = \sqrt{a_A(AO_2)} = \sqrt{914(0.102)} \,\text{m/s} = 9.66 \,\text{m/s, and}$$

$$\omega_2 = \sqrt{a_A/(AO_2)} = \sqrt{914/(0.102)} \,\text{rad/s} = 94.66 \,\text{rad/s.}$$

With reference to the acceleration diagram and configuration diagram, $\vec{\omega}_2 = -94.66\,\vec{k}$ rad/s and the direction of v_A is shown in Figure 9.4b. Similarly, with reference to the acceleration diagram and the configuration diagram, one obtains

$$v_{B3/A} = \sqrt{(a_{B3/A})_n(AB_3)} = \sqrt{99.1(0.356)} \,\text{m/s} = 5.94 \,\text{m/s}$$

and its direction is perpendicular to link 3. Thus, the velocity diagram can be constructed as shown in Figure 9.4b with a scale of 1:100.

The velocity $\vec{v}_{g3}$ represents by the line og_3 where the point g_3 is located by using the image method such that

$$\frac{ag_3}{b_{B_3}g_3} = \frac{Ag_3}{Bg_3}.$$

This completes the velocity diagram construction.

From the velocity diagram, one has the magnitudes of the velocities

$$v_{B_3} = 7.60 \,\text{m/s,} \quad v_{g_3} = 7.8 \,\text{m/s.}$$

With the co-ordinates chosen as shown in Figure 9.4a, the velocities

$$\vec{v}_{B_3} = 7.60\,\vec{\imath} \,\text{m/s,} \quad \vec{v}_{g_3} = 7.60\,\vec{\imath} + 1.66\,\vec{\jmath} \,\text{m/s.}$$

The angular velocities

$$\vec{\omega}_3 = -\sqrt{\frac{(a_{B_3/A})_n}{AB_3}}\,\vec{k}\,\frac{\text{rad}}{\text{s}} = -\sqrt{\frac{99.1}{0.356}}\,\vec{k}\,\frac{\text{rad}}{\text{s}} = -16.68\,\vec{k} \,\text{rad/s.}$$

$$\vec{\omega}_4 = \vec{\omega}_3 = -16.68\,\vec{k} \,\text{rad/s.}$$

(c) *Making use of acceleration diagram*

While the acceleration diagram is given in Figure 9.4a, one still requires to find the acceleration $\vec{a}_{g3}$ which is to be applied in the virtual work method. The acceleration $\vec{a}_{g3}$ represents by the line $o'g'_3$ where the point g'_3 is located by using the image method such that

$$\frac{a'g'_3}{b'_3 g'_3} = \frac{Ag_3}{Bg_3}.$$

With reference to the chosen co-ordinate system and Figure 9.4c where a scale of 1:10,000 was used, the vector

$$\vec{a}_{g3} = 487\,\vec{\imath} - 390\,\vec{\jmath} \ \text{m/s}^2.$$

Recall from Part (b) above, $\vec{\alpha}_2 = 0$. To calculate the angular acceleration of links 3 and 4, one makes use of the given acceleration diagram so that

$$\vec{\alpha}_3 = \sqrt{\frac{(a_{B3/A})_t}{AB_3}}\,\vec{k}\ \text{rad/s}^2 = \sqrt{\frac{468}{0.356}}\,\vec{k}\ \text{rad/s}^2 = 1315\vec{k}\ \text{rad/s}^2 = \vec{\alpha}_4.$$

(d) *Application of virtual work method*

Collect all the terms required for this part of the solution,

$$\vec{F}_s = 900\,\vec{\imath}\ \text{N},\ \vec{\omega}_2 = -94.66\vec{k}\ \text{rad/s},\ \vec{\omega}_3 = \vec{\omega}_4 = -16.68\vec{k}\ \text{rad/s},$$
$$\vec{\alpha}_2 = 0,\ \vec{\alpha}_3 = \vec{\alpha}_4 = 1315\vec{k}\ \text{rad/s}^2,$$
$$\vec{v}_{B3} = \vec{v}_{g4} = 7.60\,\vec{\imath}\ \text{m/s},\ \vec{v}_{g3} = 7.60\,\vec{\imath} + 1.66\,\vec{\jmath}\ \text{m/s},$$
$$\vec{a}_{g3} = 487\,\vec{\imath} - 390\,\vec{\jmath}\ \text{m/s}^2,$$

and recall $\vec{\imath} \bullet \vec{\jmath} = 0$, $\vec{\jmath} \bullet \vec{\jmath} = 1$, and so on.

Applying Equation (9.6), one has

$$\vec{F}_{03} \bullet \vec{v}_{g3} + \vec{F}_s \bullet \vec{v}_{B3} + \vec{T}_2 \bullet \vec{\omega}_2 + \vec{T}_{03} \bullet \vec{\omega}_3 + \vec{T}_{04} \bullet \vec{\omega}_4 = 0. \qquad (9.10)$$

Now, consider every term in this equation in which

$$\vec{F}_{03} \bullet \vec{v}_{g3} = -m_3 \vec{a}_{g3} \bullet \vec{v}_{g3}$$
$$= -(6.80)(487\vec{\imath} - 390\vec{\jmath}) \bullet (7.60\vec{\imath} + 1.66\vec{\jmath})\,\text{Nm/s}$$
$$= -20765.84\,\text{Nm/s}.$$
$$\vec{F}_s \bullet \vec{v}_{B3} = (900\vec{\imath}) \bullet \vec{v}_{B3}$$
$$\vec{F}_s \bullet \vec{v}_{B3} = (900\vec{\imath}) \bullet (7.60\vec{\imath})\,\text{Nm/s}$$
$$= 6840\,\text{Nm/s}.$$
$$\vec{T}_2 \bullet \vec{\omega}_2 = (T_2\vec{k}) \bullet (-94.66\vec{k}) = -94.66T_2.$$
$$\vec{T}_{03} \bullet \vec{\omega}_3 = -I_3\vec{\alpha}_3 \bullet \vec{\omega}_3 = -0.068(1315\vec{k}) \bullet (-16.68\vec{k})\,\text{Nm/s}$$
$$= 1491.5\,\text{Nm/s}.$$
$$\vec{T}_{04} \bullet \vec{\omega}_4 = -I_4\vec{\alpha}_4 \bullet \vec{\omega}_4 = -0.108(1315\vec{k}) \bullet (-16.68\vec{k})\,\text{Nm/s}$$
$$\vec{T}_{04} \bullet \vec{\omega}_4 = 2368.85\,\text{Nm/s}.$$

Substituting all the above results into Equation (9.10), one obtains

$$-20765.84 + 6840 - 94.66T_2 + 1491.5 + 2368.85\,\text{Nm/s} = 0.$$

This gives $T_2 = -106.33$ Nm/s or $T_2 = 106.33$ Nm/s, cw.

Example 9.3

The rotor of a jet airplane engine is supported by two bearings as shown in Figure 9.5a. The rotor assembly including compressor, turbine, and shaft has a mass of 1,000 kg and a radius of gyration of 250 mm. Determine the maximum bearing force as the airplane undergoes a pullout on a circular curve whose radius is 2,000 m at a constant airplane speed of 1,000 km/h and the engine rotor speed of 10,000 rpm. In the solution, include the gyroscopic effect and the effect of centrifugal force due to the pullout.

Solution:
Given data:

$$\begin{aligned}
&\text{mass rotor assembly} && m = 1000 \text{ kg,} \\
&\text{radius of gyration} && \gamma = 250 \text{ mm,} \\
&\text{radius of path} && R = 2000 \text{ m,} \\
&\text{velocity of airplane} && v = \frac{(1000\,\frac{\text{km}}{\text{h}})(1000\,\frac{\text{m}}{\text{km}})}{(3600\,\frac{\text{s}}{\text{h}})} = 277.78 \text{ m/s, and} \\
&\text{rotor rotating speed} && \omega = \frac{2\pi(10{,}000)}{60}\,\frac{\text{rad}}{\text{s}} = 1047.20\,\frac{\text{rad}}{\text{s}}.
\end{aligned}$$

The direction of rotation is ccw, viewed from the rhs, as indicated in Figure 9.5a.

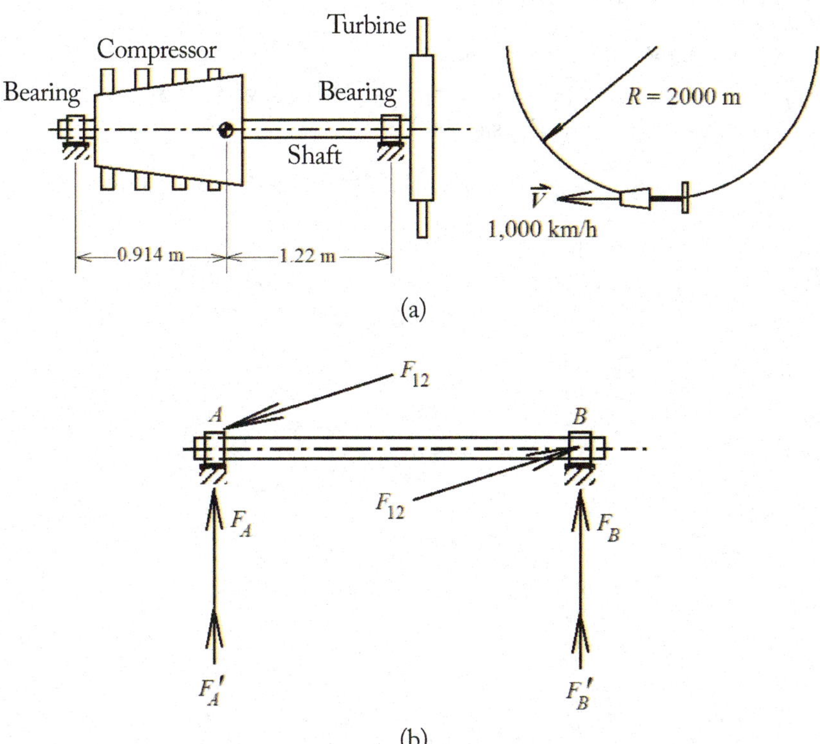

Figure 9.5: Rotor assembly and motion: (a) rotor assembly and motion path, and (b) forces at bearings.

Gyroscopic force:

$T = I\omega\omega_p$, since $I = m\gamma^2 = 1000(250/1000)^2$ kg m^2 = 62.5 kg m^2,

$$\omega_p = v/R = 277.78/2{,}000 \text{ rad/s} = 0.1389 \text{ rad/s}.$$

Therefore, $T = I\omega\omega_p = 62.5(1047.20)(0.1389)$ Nm = 9090.35 Nm. The bearing force on the shaft in the horizontal plane is given by

$$F_{12} = T/(AB) = 9090.35/2.134 \text{ N} = 4259.77 \text{ N}.$$

Centrifugal force:

The centrifugal force $F = m(\frac{v^2}{R}) = 1000(\frac{277.78^2}{2000})$ N = 38,580.86 N. Force on bearing A on shaft upward $F_A = \frac{1.22(38,580.86)}{AB}$ N = 22056.54 N. Since the force F is acting downward in the vertical plane (force on rotor and shaft on bearings), $F_B = F - F_A$, upward in the vertical plane.

Therefore, $F_B = 16524.32$ N, upward in the vertical plane.

Bearing forces due to weight of the rotor assembly:
$F'_A = \frac{1.22(1000)(9.81)}{AB}$ N $= \frac{1.22(1000)(9.81)}{2.134}$ N $= 5608.34$ N, upward in the vertical plane (force of bearing A on shaft). Bearing force F'_B is given by

$$F'_B = mg - F'_A = 1000(9.81) - 5608.34 \text{ N} = 4201.66 \text{ N},$$

upward (force of bearing B on shaft).

Therefore, $F_A + F'_A = 22056.54 + 5608.34$ N $= 27{,}664.88$ N, upward.
Similarly, $F_B + F'_B = 16524.32 + 4201.66$ N $= 20{,}725.98$ N, upward.

Total bearing forces:
Total bearing force at A,

$$F_{TA} = \sqrt{(F_A + F'_A)^2 + F_{12}^2} \text{ N}$$

$$= \sqrt{27{,}664.88^2 + (4259.77)^2} \text{ N}$$

$$= 27{,}990.91 \text{ N}.$$

Similarly, total bearing force at B,

$$F_{TB} = \sqrt{(F_B + F'_B)^2 + F_{12}^2} \text{ N}$$

$$= \sqrt{20{,}72598^2 + (4259.77)^2} \text{ N}$$

$$= 21{,}159.20 \text{ N}.$$

The forces and directions are illustrated in Figure 9.5b.

9.5   EXERCISES

9.1. By applying the virtual work method, calculate the shaft torque $\vec{T}_2$ which must be applied to link 2 of the four-bar mechanism in Figure 9.6 to hold the linkage in equilibrium at $\omega_2 = 250$ rad/s cw. The system parameters are:

$$O_2A = 76.2 \text{ mm}, \quad m_2 = 2.0 \text{ kg}, \quad I_2 = 0.002 \text{ kg m}^2,$$
$$AB = 305 \text{ mm}, \quad m_3 = 4.50 \text{ kg}, \quad I_3 = 0.030 \text{ kg m}^2,$$
$$O_2g_2 = 25.4 \text{ mm}, \quad m_4 = 7.0 \text{ kg}, \quad I_4 = 0.020 \text{ kg m}^2,$$
$$Ag_3 = 102 \text{ mm}, \quad O_4B = 152 \text{ mm}, \quad O_4g_4 = 102 \text{ mm}.$$

9.2. By applying the virtual work method, determine the crankshaft torque $\vec{T}_2$ of the mechanism shown in Figure 9.7 to hold the linkage in equilibrium $\omega_2 = 150$ rad/s ccw. The system parameters are:

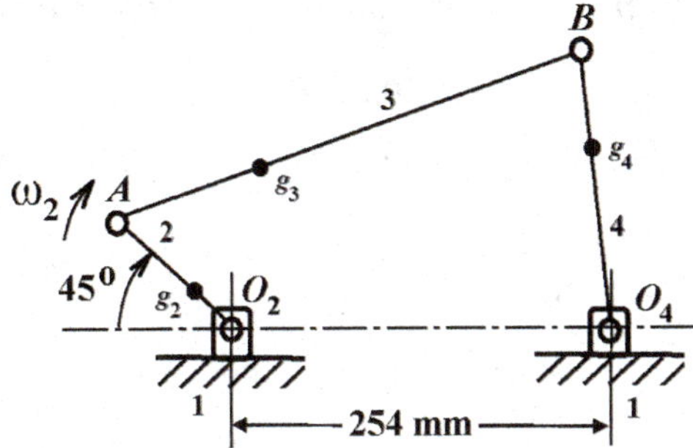

Figure 9.6: Four-bar mechanism.

$$O_2A = 76.2 \text{ mm}, \quad m_2 = 1.90 \text{ kg}, \quad I_2 = 0.003 \text{ kg m}^2,$$
$$AB = 178 \text{ mm}, \quad m_3 = 3.50 \text{ kg}, \quad I_3 = 0.030 \text{ kg m}^2,$$
$$Ag_3 = 76.2 \text{ mm}, \quad m_4 = 9.0 \text{ kg}, \quad I_4 = 0.015 \text{ kg m}^2,$$

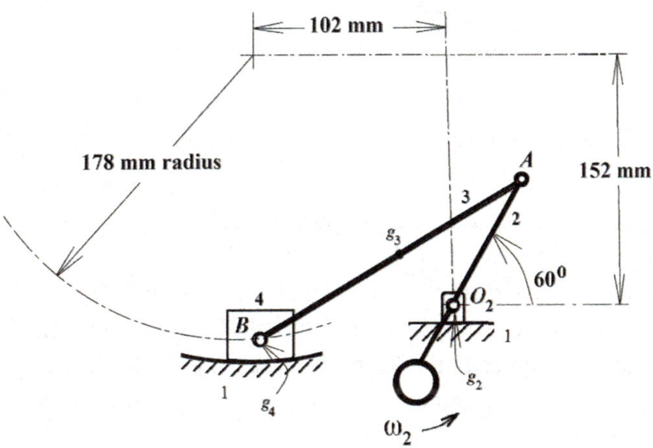

Figure 9.7: Mechanism with slider on circular ground.

9.3. For the two-cylinder engine shown in Figure 9.8 determine the crankshaft torque $\vec{T_2}$ necessary to hold the linkage in equilibrium by applying the method of virtual work. The geometrical and material properties of the engine are:

$$\text{crank, } O_2A = 51 \text{ mm, connecting rod, } AB = AC = 204 \text{ mm,}$$

distance from crank center to center of gravity of connecting rod,

$$Ag_3 = Ag_4 = 50.8 \, \text{mm}, \quad m_3 = m_4 = 1.30 \, \text{kg},$$

$$m_5 = m_6 = 1.00 \, \text{kg}, \quad I_3 = I_4 = 0.012 \, \text{kg} \, \text{m}^2.$$

Note that the mass of the connecting rod, for example m_3, is the sum of two parts which are the mass of crank pin, $m_{3c} = 1.00$ kg and the mass at wrist pin, $m_{3w} = 0.30$ kg.

The gas forces along the axes of cylinders 1 and 2 are, respectively,

$$P_1 = 8,000 \, \text{N}, \quad P_2 = 1,780 \, \text{N}.$$

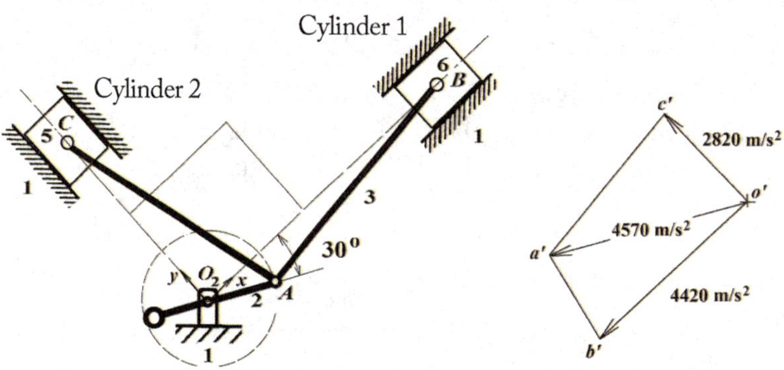

Figure 9.8: Two-cylinder engine and its acceleration diagram.

9.4. The angular velocity of a 1,500 kg space capsule is $\vec{\omega}_c = 0.02\vec{\imath} + 0.10\vec{\jmath}$ rad/s when two small jets are activated at A and B, each in a direction parallel to the z-axis. This capsule is shown in Figure 9.9. Knowing that the radii of gyration of the capsule are $\gamma_x = \gamma_z = 1.00$ m and $\gamma_y = 1.20$ m, and that each of jet generates a trust of 30 N, determine (a) the required operation time of each jet if the angular velocity of the capsule is to be reduced to zero, and (b) the resulting change in the velocity of the mass center G.

9.5. The conceptual model of a type of aircraft turn indicator is shown in Figure 9.10. Springs AC and BD are initially stretched and exert equal vertical forces at A and B when the path of the airplane is straight. If the rotating disk has a mass of 0.10 kg and spins at 9,000 rpm, determine the angle through which the yoke rotates when the aircraft executes a horizontal turn of radius 800 m at a speed of 600 km/h. Given that each spring constant is 1,000 N/m.

9.6. An electric motor drives a railway car. The mass of the armature of the motor is 280 kg and the radius of gyration of the armature is 150 mm. The axis of the armature runs

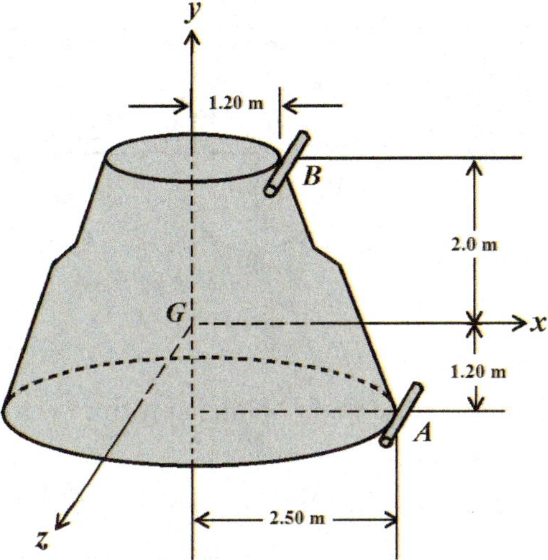

Figure 9.9: Space capsule with two small jets.

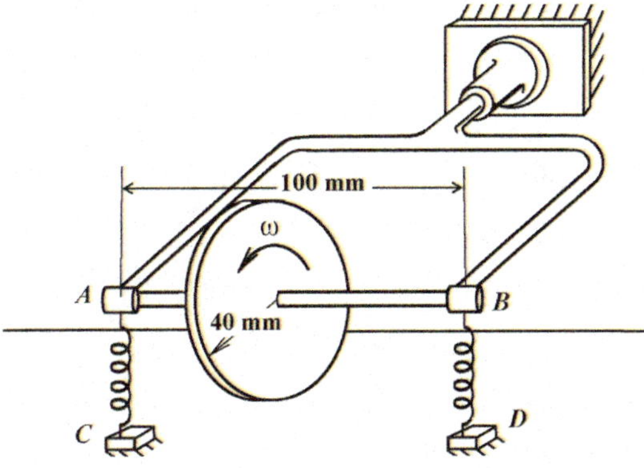

Figure 9.10: Conceptual model of an aircraft turn indicator.

parallel to the axles of the car. The armature rotates in the opposite sense to that of the car wheels, and it makes 8 revolutions for one revolution of the car wheels, each of which has a diameter of 1.0 m. The distance between the armature bearings is 600 mm. If the car is moving forward with a speed of 8 m/s around a curve to the right, determine the total force each bearing exerts on the armature shaft, assuming the radius of the curve is 50 m. Before the computation make a sketch of the armature shaft looking at it from the rear of the car. Note that the sketch should be similar to that shown in Figure 9.11.

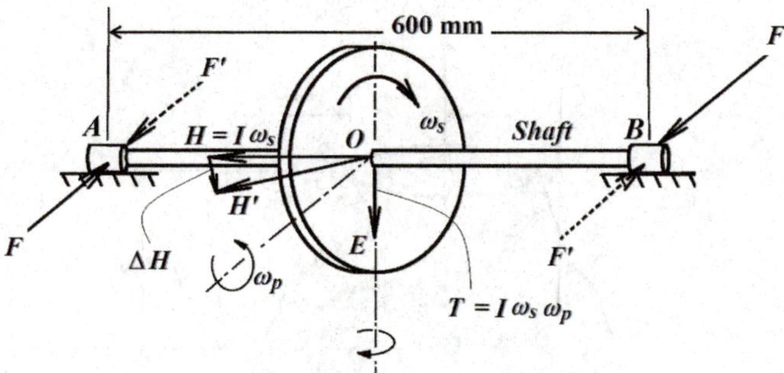

Figure 9.11: Similar sketch of armature shaft looking at it from rear of car.

REFERENCES

[1] Beer, F. P., Johnston, Jr., E. R., and Cornwell, P. J., *Vector Mechanics for Engineers: Dynamics*, 10th edition, New York, McGraw-Hill, 2013. 162

[2] Mabie, H. H. and Ocvirk, F. W., *Mechanisms and Dynamics of Machinery*, 3rd edition-SI Version, New York, John Wiley & Sons, 1978. 168

Author's Biography

CHO W. S. TO

Dr. To obtained his doctoral degree in sound and vibration studies from the University of Southampton in April 1980. He is currently a fellow of the American Society of Mechanical Engineers (ASME) and a professor emeritus in the Department of Mechanical and Materials Engineering at the University of Nebraska (UNL). Prior to joining UNL he was a professor (1994-96) and an associate professor (1986-94) at the University of Western Ontario, Canada. He was an associate professor (1985-86) and an assistant professor (1982-85) at the University of Calgary. Between 1982 and 1992 he was a University Research Fellow of the Natural Sciences and Engineering Research Council (NSERC), Canada. He served as chair of the ASME Finite Element Techniques and Computational Technologies Technical Committee. In addition to being a member of the editorial boards for several refereed international journals, Dr. To currently serves as an associate editor of the *International Journal of Mechanics*.

His main academic interests are in nonlinear stochastic structural dynamics, nonlinear finite element analysis with particular reference to laminated composite plates and shells, nonlinear dynamics and control, machinery noise and vibration diagnostics, and mechanics of carbon nano-tubes. He is the author of *Introduction to Dynamics and Control in Mechanical Engineering Systems* (2016, ASME and Wiley), *Stochastic Structural Dynamics: Application of Finite Element Methods* (2014, Wiley), *Nonlinear Random Vibration: Analytical Techniques and Applications* (Second Edition, 2012, Taylor and Francis/CRC Press), and *Nonlinear Random Vibration: Computational Methods* (2010, Zip Publishing, Columbus, Ohio).

Index

9 781681 731759